The Secret Life of
the Mind

The Secret Life of the Mind

How Your Brain Thinks, Feels, and Decides

MARIANO SIGMAN

To Milo and Noah

Little, Brown and Company
Hachette Book Group
1290 Avenue of the Americas, New York, NY 10104
littlebrown.com

Originally published in Buenos Aires and Barcelona by Penguin Random House Grupo Editorial, S.A. in 2015
First North American Edition: June 2017

Little, Brown and Company is a division of Hachette Book Group, Inc. The Little, Brown name and logo are trademarks of Hachette Book Group, Inc.

The publisher is not responsible for websites (or their content) that are not owned by the publisher.

The Hachette Speakers Bureau provides a wide range of authors for speaking events. To find out more, go to hachettespeakersbureau.com or call (866) 376-6591.

ISBN 978-0-316-54962-2
LCCN 2017939488

10 9 8 7 6 5 4 3 2 1

LSC-C

Printed in the United States of America

CONTENTS

2: The fuzzy borders of identity   47
What defines our choices and allows us to trust other
people and our own decisions?

3: The machine that constructs reality   99
How does consciousness emerge in the brain and
how are we governed by our unconscious?

INTRODUCTION

I like to think of science as a ship that takes us to unknown places, the remotest parts of the universe, into the inner workings of light and the tiniest molecules of life. This ship has instruments, telescopes and microscopes, which make visible what was once invisible. But science is also the route itself, the binnacle, the chart leading us towards the unknown.

My voyage over the last twenty years, between New York, Paris and Buenos Aires, has been into the innermost parts of the human brain, an organ composed of countless neurons that codify perception, reason, emotions, dreams and language.

The goal of this book is to discover our mind in order to understand ourselves more deeply, even in the tiniest recesses that make up who we are. We will look at how we form ideas during our first days of life, how we give shape to our fundamental decisions, how we dream and how we imagine, why we feel certain emotions, how the brain transforms and how who we are changes along with it.

Throughout these pages we will see the brain from a distance. We will go where thought begins to take shape. And it is there where psychology meets neuroscience. This is the ocean in which many people of varied disciplines have sailed, including biologists, physicists, mathematicians, psychologists, anthropologists, linguists, philosophers, doctors. As well as chefs, magicians, musicians, chess masters, writers, artists. This book is a result of that amalgam.

The first chapter is a journey to the land of childhood. We will see that the brain is already prepared for language long before we begin to speak, that bilingualism helps us to think, and that early on we form notions of what is good and what is fair, and about cooperation and competition, that later affect how we relate to ourselves and others. This early intuitive thinking leaves lasting traces on the way we reason and decide.

In the second chapter we will explore what defines the blurry, fine line between what we are willing to do and what we aren't. Those decisions that make us who we are. How do reason and feelings work together in social and emotional decisions? What makes us trust others and ourselves? We will discover that small differences in decision-making brain circuits can drastically change our way of deciding, from the simplest decisions to the most profound and sophisticated ones that define us as social beings.

The third and fourth chapters travel into the most mysterious aspect of thought and the human brain – consciousness – through an unprecedented meeting between Freud and the latest neuroscience. What is the unconscious and how does it control us? We will see that we can read and decipher thoughts by decoding patterns of brain activity, even in vegetative patients who have no other way to express themselves. And who is it that awakens when consciousness awakens? We will see the first sketches of how we can now record our dreams and visualize them within some sort of oneiric planetarium, and explore the fauna of different states of consciousness, like lucid dreams and thinking under the effects of marijuana or hallucinogenic drugs.

The last two chapters cover questions of how the brain learns in different circumstances, from everyday life to formal education. For example, is it true that learning a new language is much harder for an adult than for a child? We will take a journey into the history of learning, looking at effort and ability, the drastic transformation that takes place in the brain when we learn to read, and the brain's predisposition to change. This book outlines how all this knowledge can be used responsibly to improve the largest collective experiment in the history of humanity: school.

The Secret Life of the Mind is a summary of neuroscience from the perspective of my own experience. I look at neuroscience as a way to help us communicate with each other. From this perspective, neuroscience is another tool in humanity's ancestral search to express – sometimes rudimentarily – the shades, colours and nuances of what we feel and what we think in order to be comprehensible to others and, of course, to ourselves.

CHAPTER ONE

The origin of thought

How do babies think and communicate, and how can we understand them better?

Of all the places we travel throughout our lifetimes, the most extraordinary is certainly the land of childhood: a territory that, looked back on by the adult, becomes a simple, naive, colourful, dreamlike, playful and vulnerable space.

It's odd. We were all once citizens of that country, yet it is hard to remember and reconstruct it without dusting off photos in which, from a distance, we see ourselves in the third person, as if that child were someone else and not us in a different time.

How did we think and conceive of the world before learning the words to describe it? And, while we are at it, how did we discover those words without a dictionary to define them? How is it possible that before three years of age, in a period of utter immaturity in terms of formal reasoning, we were able to discover the ins and outs of grammar and syntax?

Here we will sketch out that journey, from the day we entered the world to the point where our language and thought resemble what we employ today as adults. The trajectory makes use of diverse vehicles, methods and tools. It intermingles reconstructions of thought from our gazes, gestures and words, along with the minute inspection of the brain that makes us who we are.

We will see that, from the day we are born, we are already able to form abstract, sophisticated representations. Although it sounds far-fetched, babies have notions of mathematics, language, morality, and even scientific and social reasoning. This creates a repertoire of innate intuitions that structure what they will learn – what we all learned – in social, educational and family spaces, over the years of childhood.

We will also discover that cognitive development is not the mere acquisition of new abilities and knowledge. Quite the contrary, it often consists in undoing habits that impede children from demonstrating what they already know. On occasion, and despite it being a counterintuitive idea, the challenge facing children is not acquiring new concepts but rather learning to manage those they already possess.

I have observed that we, as adults, often draw babies poorly because we don't realize that their body proportions are completely different from ours. Their arms, for example, are barely the size of their heads. Our difficulty in seeing them as they are serves as a morphological metaphor for understanding what is most difficult to sense in the cognitive sphere: babies are not miniature adults.

In general, for simplicity and convenience, we speak of *children* in the third person, which erroneously assumes a distance, as if we were talking about something that is not us. Since this book's intention is to travel to the innermost recesses of our brain, this first excursion, to the child we once were, will be in the first person in order to delve into how we thought, felt and represented the world in those days we can no longer recall, simply because that part of our experience has been relegated to oblivion.

The genesis of concepts

In the late seventeenth century, an Irish philosopher, William Molyneux, suggested the following mental experiment to his friend John Locke:

Suppose a man born blind, and now adult, and taught by his touch to distinguish between a cube and a sphere [. . .] Suppose then the cube and the sphere placed on a table, and the blind man made to see: query, Whether by his sight, before he touched them, he could now distinguish and tell which is the globe, which the cube?

Could he? In the years that I have been asking this question I've found that the vast majority of people believe that the answer is no. That the virgin visual experience needs to be linked to what is already known through touch. Which is to say, that a person would need to feel and see a sphere at the same time in order to discover that the gentle, smooth curve perceived by the fingertips corresponds to the image of the sphere.

Others, the minority, believe that the previous tactile experience creates a visual mould. And that, as a result, the blind man would be able to distinguish the sphere from the cube as soon as he could see.

John Locke, like most people, thought that a blind man would have to learn how to see. Only by seeing and touching an object at the same time would he discover that those sensations are related, requiring a translation exercise in which each sensory mode is a different language, and abstract thought is some sort of dictionary that links the *tactile words* with the *visualized words*.

For Locke and his empiricist followers, the brain of a newborn is a blank page, a *tabula rasa* ready to be written on. As such, experience goes about sculpting and transforming it, and concepts are born only when they acquire a name. Cognitive development begins on the surface with sensory experience, and, then, with the development of language, it acquires the nuances that explain the deeper and more sophisticated aspects of human thought: love, religion, morality, friendship and democracy.

Empiricism is based on a natural intuition. It is not surprising, then, that it has been so successful and that it dominated the philosophy of the mind from the seventeenth century to the time of the great Swiss psychologist Jean Piaget. However, reality is not always intuitive: the brain of a newborn is not a *tabula rasa*. Quite the contrary. We already come into the world as conceptualizing machines.

The typical café discussion reasoning comes up hard against reality in a simple experiment carried out by a psychologist, Andrew Meltzoff, in which he tested a version of Molyneux's question in order to refute empirical intuition. Instead of using a sphere and a cube, he used two dummies: one smooth and rounded and the other more bumpy, with nubs. The method is simple. In complete darkness, babies had one of the two pacifiers in their mouths. Later, the pacifiers are placed on a table and the light is turned on. And the babies looked more at the pacifier they'd had in their mouths, showing that they recognize it.

The experiment is very simple and destroys a myth that had persisted over more than three hundred years. It shows that a newborn with only tactile experience – contact with the mouth, since at that age tactile exploration is primarily oral as opposed to manual – of an object has already conceived a representation of how it looks. This contrasts with what parents typically perceive: that newborn babies' gazes often seem to be lost in the distance and disconnected from reality. As we will see later, the mental life of children is actually much richer and more sophisticated than we can intuit based on their inability to communicate it.

Atrophied and persistent synaesthesias

Meltzoff's experiment gives – against all intuition – an affirmative response to Molyneux's question: newborn babies can recognize by sight two objects that they have only touched. Does the same thing happen with a blind adult who begins to see? The answer to this question only recently became possible once surgeries were able to reverse the thick cataracts that cause congenital blindness.

The first actual materialization of Molyneux's mental experiment was done by the Italian ophthalmologist Alberto Valvo. John Locke's prophecy was correct; for a çongenitally blind person, gaining sight was nothing like the dream they had longed for. This was what one of the patients said after the surgery that allowed him to see:

> I had the feeling that I had started a new life, but there were moments when I felt depressed and disheartened, when I realized how difficult it was to *understand* the visual world. [. . .] In fact, I see groups of lights and shadows around me [. . .] like a mosaic of shifting sensations whose *meaning* I don't understand. [. . .] At night, I like the darkness. I had to die as a blind person in order to be reborn as a seeing person.

This patient felt so challenged by suddenly gaining sight because while his eyes had been 'opened' by the surgery, he still had to learn to see. It was a big and tiresome effort to put together the new visual experience with the conceptual world he had built through his senses of hearing and touch. Meltzoff proved that the human brain has the ability to establish spontaneous correspondences between sensory modalities. And Valvo showed that this ability atrophies when in disuse over the course of a blind life.

On the contrary, when we experience different sensory modalities, some correspondences between them consolidate spontaneously over time. To prove this, my friend and colleague Edward Hubbard, along with Vaidyanathan Ramachandran, created the two shapes that we see here. One is Kiki and the other is Bouba. The question is: which is which?

Almost everyone answers that the one on the left is Bouba and the one on the right is Kiki. It seems obvious, as if it couldn't be any other way. Yet there is something strange in that correspondence; it's like saying someone *looks like a Carlos*. The explanation for this is that when we pronounce the vowels /o/y/u/, our lips form a wide circle, which corresponds to the roundness of Bouba. And when saying the /k/, or /i/, the back part of the tongue rises and touches the palate in a very angular configuration. So the pointy shape naturally corresponds with the name Kiki.

These bridges often have a cultural basis, forged by language. For example, most of the world thinks that the past is behind us and the future is forward. But that is arbitrary. For example, the Aymara, a people from the Andean region of South America, conceive of the association between time and space differently. In Aymara, the word 'nayra' means past but also means in front, in view. And the word 'quipa', which means future, also indicates behind. Which is to say that in the Aymaran language the past is ahead and the future behind. We know that this reflects their way of thinking, because they also express that relationship with their bodies. The Aymara extend their arms backwards to refer to the future and forwards to allude to the past. While on the face of it this may seem strange, when they explain it, it seems so reasonable that we feel tempted to change our own way of envisioning it; they say that the past is the only thing we know – what our eyes see – and therefore it is in front of us. The future is the unknown – what our eyes do not know – and thus it is at our backs. The Aymara walk backwards through their timeline. Thus, the uncertain, unknown future is behind and gradually comes into view as it becomes the past.

We designed an atypical experiment, with the linguist Marco Trevisan and the musician Bruno Mesz, in order to find out whether there is a natural correspondence between music and taste. The experiment brought together musicians, chefs and neuroscientists. The musicians were asked to improvise on the piano, based on the four canonical flavours: sweet, salty, sour and bitter. Of course, coming from different musical schools and styles (jazz, rock, classical, etc.) each one of them had their own distinctive interpretation. But within that wide variety we found that each taste inspired consistent patterns: the bitter corresponded with deep, continuous tones; the salty with notes that were far apart (*staccato*); the sour with very high-pitched, dissonant melodies; and the sweet with consonant, slow and gentle music. In this way we were able to *salt* 'Isn't She Lovely' by Stevie Wonder and to make a *sour* version of *The White Album* by the Beatles.

The mirror between perception and action

Our representation of time is random and fickle. The phrase 'Christmas is fast approaching' is strange. Approaching from where? Does it come from the south, the north, the west? Actually, Christmas isn't located anywhere. It is in time. This phrase, or the analogous one, 'we're getting close to the end of the year', reveals something of how our minds organize our thoughts. We do it in our bodies. Which is why we talk of the *head* of government, of someone's right-*hand* man, the *armpit* of the world and many other metaphors* that reflect how we organize thought in a template defined by our own bodies. And because of that, when we think of others' actions, we do so by acting them out ourselves, speaking others' words in our own voice, yawning someone else's yawn and laughing someone else's laugh. You can do a simple experiment at home to test out this mechanism. During a conversation, cross your arms. It's very likely that the person you are speaking to will do the same. You can take it further with bolder gestures, like touching your head, or scratching yourself, or stretching. The probability that the other person will imitate you is high.

---

* The head on a beer, the eye of a storm, the arms of a river, a zipper's teeth, the open veins of Latin America and this footnote.

This mechanism depends on a cerebral system made up of *mirror neurons*. Each one of these neurons codifies specific gestures, like moving an arm or opening up a hand, but it does so whether or not the action is our own or someone else's. Just as the brain has a mechanism that spontaneously amalgamates information from different sensory modes, the mirror system allows – also spontaneously – our actions and others' actions to be brought together. Lifting your arm and watching someone else do it are very different processes, since one is done by you and the other is not. As such, one is visual and the other is motor. However, from a conceptual standpoint, they are quite similar. They both correspond to the same gesture in the abstract world.

And now after understanding how we adults merge sensory modalities in music, in shapes and sounds and in language, and how we bring together perception and action, we go back to the infant mind, specifically to ask whether the mirror system is learned or whether it is innate. Can newborns understand that their own actions correspond to the observation of another person's? Meltzoff also tested this out, to put an end to the empirical idea that considers the brain a *tabula rasa*.

Meltzoff proposed another experiment, in which he made three different types of face at a baby: sticking out his tongue, opening his mouth, and pursing his lips as if he were about to give the child a kiss. He observed that the baby tended to repeat each of his gestures. The imitation wasn't exact or synchronized; the mirror is not a perfect one. But, on average, it was much more likely that the baby would replicate the gesture he or she observed than make one of the other two. Which is to say that newborns are capable of associating observed actions with their own, although the imitation is not as precise as it will later become when language is introduced.

Meltzoff's two discoveries – the associations between our actions and those of others, and between varying sensory modalities – were published in 1977 and 1979. By 1980, the empirical dogma was almost completely dismantled. In order to deal it a final death blow, there was one last mystery to be solved: Piaget's mistake.*

Piaget's mistake!

One of the loveliest experiments done by the renowned Swiss psychologist Jean Piaget is the one called *A-not-B*. The first part goes like this: there are two napkins on a table, one on each side. A ten-month-old baby is shown an object, then it is covered with the first napkin (called 'A'). The baby finds it without difficulty or hesitation.

Behind this seemingly simple task is a cognitive feat known as object permanence: in order to find the object there must be a reasoning that goes beyond what is on the surface of the senses. The object did not disappear. It is merely hidden. A baby that is to be able to comprehend this must have a view of the world in which things

---

* Throughout the book we will expose 'mistakes' in the history of psychology, science, and the philosophy of the mind. Many of these 'mistakes' reflect intuitions and, therefore, are replicated in each of our own histories. They are myths that persist beyond evidence to the contrary because they are based on natural, intuitive reasoning. As obvious as it may seem, I want to clarify that when I mention the mistakes made by great thinkers I do so from the privileged perspective of someone who has access to facts they did not, in other words, looking back – or forward – at the past. It is the difference between analysing a game and playing it, or playing fantasy baseball. I am working with the premise that science, and almost any human conjecture, is always approximate and is constantly being revised. Talking about Piaget's mistake is, from my point of view, a sort of an ode to his work, an acknowledgement of his ideas, which, while not always correct, were landmarks in the history of knowledge. As Isaac Newton said: 'If I have seen further, it is by standing on the shoulders of giants.' This is a version of the history of knowledge that is more realistic and less celebrated than the story of the apple hitting his head and giving him sudden inspiration. It goes without saying that this book is also a homage to all my great predecessors, whose hits and misses cemented the road that so many of us now travel along.

do not cease to exist when we no longer see them. That, of course, is abstract.*

The second part of the experiment begins in exactly the same way. The same ten-month-old baby is shown an object, which is then covered up by napkin 'A'. But then, and before the baby does anything, the person running the experiment moves the object to underneath the other napkin (called 'B'), making sure that the baby sees the switch. And here is where it gets weird: the baby lifts the napkin where it was first hidden, as if not having observed the switch just made in plain sight.

This error is ubiquitous. It happens in every culture, almost unfailingly, in babies about ten months of age. The experiment is striking and precise, and shows fundamental traits of our way of thinking. But Piaget's conclusion, that babies of this age still do not fully understand the abstract idea of object permanence, is erroneous.

When revisiting the experiment, decades later, the more plausible – and much more interesting – interpretation is that babies know the object has moved but cannot use that information. They have, as happens in a state of drunkenness, a very shaky control of their actions. More precisely, ten-month-old babies have not yet developed a system of inhibitory control, which is to say, the ability to prevent themselves doing something they had already planned to do. In fact, this example turns out to be the rule. We will see in the next section how certain aspects of thought that seem sophisticated and elaborated – morality or mathematics, for example – are already sketched from the day we are born. On the other hand, others that seem much more rudimentary, like halting a decision, mature gradually and steadily. To

---

* All parents play peek-a-boo, eliciting peals of laughter from their kids. That is the pleasure of discovering and understanding that objects do not disappear when we can no longer see them. Children are young scientists enjoying the process of discovering the rules of the universe.

understand how we came to know this, we need to take a closer look at the executive system, or the brain's 'control tower', which is formed by an extensive neural network distributed in the prefrontal cortex that matures slowly during childhood.

The executive system

The network in the frontal cortex that organizes the executive system defines us as social beings. Let's give a small example. When we grab a hot plate, the natural reflex would be to drop it immediately. But an adult, generally, will inhibit that reflex while quickly evaluating if there is a nearby place to set it down and avoid breaking the plate.

The executive system governs, controls and administers all these processes. It establishes plans, resolves conflicts, manages our attention focus, and inhibits some reflexes and habits. Therefore the ability to govern our actions depends on the reliability of the executive function system.* If it does not work properly, we drop the hot

---

* While I was doing my doctorate in New York I went up to Boston one day to visit the laboratory of Alvaro Pascual-Leone. At that time they were just beginning to use a tool called TMS (Transcranial Magnetic Stimulation). TMS uses a system of coils to transmit a very faint magnetic pulse that is able to activate or inhibit a region of the cerebral cortex. When I arrived, they were doing an experiment in which they temporarily deactivated the frontal cortex. I was tempted by the idea of experiencing at first hand a decrease in the functioning of the executive system and I offered myself as a subject. After they inhibited my frontal cortex – reversibly – for thirty minutes, the experiment began. I would see a letter and I had to think of words that started with it and then say them out loud a few seconds later. This waiting is controlled by the executive system. With my prefrontal cortex inhibited it was impossible for me to wait. I started naming the words compulsively, at the very moment they came into my head. I understood that I was supposed to wait before saying them, but I just couldn't. This experience in real time and in a sort of disassociation between the first person – the actor – and the third person – the observer – allowed me to directly understand the limits of what we are able to do (limits that go beyond our desire and willpower) in cognitive domains that are apparently very basic. It is extremely difficult, if you

plate, burp at the table, and gamble away all our money at the roulette wheel.

The frontal cortex is very immature in the early months of life and it develops slowly, much more so than other brain regions. Because of this, babies can only express very rudimentary versions of the executive functions.

A psychologist and neuroscientist, Adele Diamond, carried out an exhaustive and meticulous study on physiological, neurochemical and executive function development during the first year of life. She found that there is a precise relationship between some aspects of the development of the frontal cortex and babies' ability to perform Piaget's *A-not-B* task.

What impedes a baby's ability to solve this apparently simple problem? Is it that babies cannot remember the different positions the object could be hidden in? Is it that they do not understand that the object has changed place? Or is it, as Piaget suggested, that the babies do not even fully understand that the object hasn't ceased to exist when it is hidden under a napkin? By manipulating all the variables in Piaget's experiment – the number of times that babies repeat the same action, the length of time they remember the position of the object, and the way they expresses their knowledge – Diamond was able to demonstrate that the key factor impeding the solution of this task is babies' inability to inhibit the response they have already prepared. And with this, she laid the foundations of a paradigm shift: children don't always need to learn new concepts; sometimes they just need to learn how to express the ones they already know.

The secret in their eyes

So we know that ten-month-old babies cannot resist the temptation to extend their arms where they were planning to, even when they

---

do not experience it for yourself, to imagine being unable to do what almost everyone does simply and naturally.

understand that the object they wish to reach has changed location. We also know that this has to do with a quite specific immaturity of the frontal cortex in the circuits and molecules that govern inhibitory control. But how do we know if babies indeed understand that the object is hidden in a new place?

The key is in their gaze. While babies extend their arms towards the wrong place, they stare at the right place. Their gazes and their hands point to different locations. Their gaze shows that they know where it is; their hand movement shows that they cannot inhibit the mistaken reflex. They are – we are – two-headed monsters. In this case, as in so many others, the difference between children and adults is not what they know but rather how they are able to act on the basis of that knowledge.

In fact, the most effective way of figuring out what children are thinking is usually by observing their gaze.* Going with the premise that babies look more at something that surprises them, a series of games can be set up in order to discover what they can distinguish and what they cannot, and this can give answers as to their mental representations. For example, that was how it was discovered that babies, a day after being born, already have a notion of numerosity, something that previously seemed impossible to determine.

The experiment works like this. A baby is shown a series of images. Three ducks, three red squares, three blue circles, three triangles, three sticks . . . The only regularity in this sequence is an abstract, sophisticated element: they are all sets of three. Later the baby is shown two images. One has three flowers and

---

* The gaze is also one of the most revealing elements of adult thought, of how we reason and what we yearn for. It not only serves to acquire knowledge, but also speaks of who we are. But, unlike small children, adults know that their gaze can give them away. That is the origin of the shyness that is so clearly expressed in one of the most spectacular laboratories for the study of human *microsociology*: the lift.

the other four. Which do the newborns look at more? The gaze is variable, of course, but they consistently look longer at the one with four flowers. And it is not that they are looking at the image because it has more things in it. If they were shown a sequence of groups of four objects, they would later look longer at one that had a group of three. It seems they grow bored of always seeing the same number of objects and are surprised to discover an image that breaks the rule.

Liz Spelke and Véronique Izard proved that the notion of numerosity persists even when the quantities are expressed in different sensory modalities. Newborns that hear a series of three beeps expect there then to be three objects and are surprised when that is not the case. Which is to say, babies assume a correspondence of amounts between the auditory experience and the visual one, and if that abstract rule is not followed through, their gaze is more persistent. These newborns have only been out of the womb for a matter of hours yet already have the foundations of mathematics in their mental apparatus.

Development of attention

Cognitive faculties do not develop homogeneously. Some, like the ability to form concepts, are innate. Others, like the executive functions, are barely sketched in the first months of life. The most clear and concise example of this is the development of the attentional network. Attention, in cognitive neuroscience, refers to a mechanism that allows us to selectively focus on one particular aspect of information and ignore other concurrent elements.

We all sometimes – or often – struggle with attention. For example, when we are talking to someone and there is another interesting conversation going on nearby.* Out of courtesy, we want to remain focused

---

* For example, hearing our own names is a magnet for our attention.

on our interlocutor, but our hearing, gaze and thoughts generally direct themselves the other way. Here we recognize two ingredients that lead and orient attention: one endogenous, which happens from inside, through our own desire to concentrate on something, and the other exogenous, which happens due to an external stimulus. Driving a car, for example, is another situation of tension between those systems, since we want to be focused on the road but alongside it there are tempting advertisements, bright lights, beautiful landscapes – all elements that, as admen know well, set off the mechanisms of exogenous attention.

Michael Posner, one of the founding fathers of cognitive neuroscience, separated the mechanisms of attention* and found that they were made up of four elements:

(1) Endogenous orientation.
(2) Exogenous orientation.
(3) The ability to maintain attention.
(4) The ability to *disengage* it.

He also discovered that each of these processes involves different cerebral systems, which extend throughout the frontal, parietal and anterior cingulate cortices. In addition, he found that each one of these pieces of the attentional machinery develops at its own pace and not in unison.

For example, the system that allows us to orient our attention towards a new element matures much earlier than the system that allows us to disengage our attention. Therefore, voluntarily shifting our attention away from something is much more difficult than we imagine. Knowing this can be of enormous help when dealing with a child; a clear example is found in how to stop a small child's

---

* He was tired of being distracted by other people talking about Kevin Costner's films.

inconsolable crying. A trick that some parents hit upon spontaneously, and emerges naturally when one understands attention development, is not asking their offspring to just cut it out, but rather to offer another option that attracts their attention. Then, almost by magic, the inconsolable crying stops *ipso facto*. In most cases, the baby wasn't sad or in pain, but the crying was, actually, pure inertia. That this happens the same way for all children around the world is not magic or a coincidence. It reflects how we are – how we were – in that developmental period: able to draw our attention towards something when faced with an exogenous stimulus, and unable to voluntarily *disengage*.

Separating out the elements that comprise thought allows for a much more fluid relationship between people. No parent would ask a six-month-old to run, and they certainly wouldn't be frustrated when it didn't happen. In much the same way, familiarity with attentional development can avoid a parent asking a small child to do the impossible; for example, to just quit crying.

The language instinct

In addition to being connected for concept formation, a newborn's brain is also predisposed for language. That may sound odd. Is it predisposed for French, Japanese or Russian? Actually, the brain is predisposed for all languages because they all have – in the vast realm of sounds – many things in common. This was the linguist Noam Chomsky's revolutionary idea.

All languages have similar structural properties. They are organized in an auditory hierarchy of phonemes that are grouped into words, which in turn are linked to form sentences. And these sentences are organized syntactically, with a property of recursion that gives the language its wide versatility and effectiveness. On this empirical premise, Chomsky proposed that language acquisition in infancy is limited and guided by the constitutional organization of the human brain. This is another argument against the notion of the *tabula rasa*: the brain has a very precise architecture that, among other things, makes it ideal for language. Chomsky's argument has another advantage, since it explains why children can learn language so easily despite its being filled with very sophisticated and almost always implicit grammatical rules.

This idea has now been validated by many demonstrations. One of the most intriguing was presented by Jacques Mehler, who had French babies younger than five days old listen to a succession of various phrases spoken by different people, both male and female. The only thing common to all the phrases was that they were in Dutch. Every once in a while the phrases abruptly changed to Japanese. He was trying to see if that change would surprise a baby, which would show that babies are able to codify and recognize a language.

In this case, the way to measure their surprise wasn't the persistence of their gaze but the intensity with which they sucked on their dummies. Mehler found that when the language changed, the babies sucked harder – like Maggie Simpson – indicating that they perceived that something relevant or different was occurring. The key is that this did not happen when he repeated the same experiment with the sound of all the phrases reversed, like a record played backwards. That means that the babies didn't have the ability to recognize categories from just any sort of sound but rather they were specifically tuned to process languages.

We usually think that innate is the opposite of learned. Another way of looking at it is thinking of the innate as actually something learned in the slow cooker of human evolutionary history. Following this line of reasoning, since the human brain is already predisposed for language at birth, we should expect to find precursors of language in our evolutionary cousins.

This is precisely what Mehler's group proved by showing that monkeys also have auditory sensibilities attuned to language. Just like babies, tamarin monkeys reacted with the same surprise every time the language they were hearing in the experiment changed. As with babies, this was specific to language, and did not happen when phrases were played backwards.

This was a spectacular revelation, not to mention a gift for the media . . . 'Monkeys Speak Japanese' is a prime example of how to destroy an important scientific finding with a lousy headline. What this experiment proves is that languages are built upon a sensitivity of the primate brain to certain combinations of sounds This in turn may explain in part why most of us learn to understand spoken language so easily at a very young age.

Mother tongue

Our brains are prepared and predisposed for language from the day we are born. But this predisposition does not seem to materialize without social experience, without using it with other people. This conclusion comes from studies of *feral children* who grow up without any human contact. One of the most emblematic is Kaspar Hauser, magnificently portrayed in the eponymous film directed by Werner Herzog. Kaspar Hauser's story of confinement for the duration of his childhood* shows that it is very difficult to acquire language when it has not been practised early in life. The ability to speak a language, to a large extent, is learned in a community. If a child grows up in complete

---

* Kaspar Hauser was a German boy who claimed he had been raised in total isolation in a dark basement. He was found in 1828 wandering the streets of Nuremberg unable to speak but a few words in German. It is believed that he was sixteen years old at that time. This case, as with similar examples of feral children, remains controversial to this day, and many cases have been shown to be poorly documented and fuelled by literary narratives. Hence, the hard conclusion that language *cannot* be learned if it is not practised early in life may have to be softened. (See Adriana Benzaquén's book, *Encounters with Wild Children: Temptation and Disappointment in the Study of Human Nature* (McGill–Queen's Press, 2006)).

isolation from others, his or her ability to learn a language is largely impaired. Herzog's film is, in many ways, a portrait of that tragedy.

The brain's predisposition for a universal language becomes fine-tuned by contact with others, acquiring new knowledge (grammatical rules, words, phonemes) or unlearning differences that are irrelevant to one's mother tongue.

The specialization of language happens first with phonemes. For example, in Spanish there are five vowel sounds, while in French, depending on the dialect, there are up to seventeen (including four nasal vowel sounds). Non-French speakers often do not perceive the difference between some of these vowel sounds. For instance, native Spanish speakers typically do not distinguish the difference between the sounds of the French words *cou* (pronounced [ku]) and *cul* (pronounced [ky]) which may lead to some anatomical misunderstanding since *cou* means neck and *cul* means bum. Vowels that they perceive as [u] in both cases sound completely different for a French speaker, as much so as an 'e' and an 'a' for Spanish speakers. But the most interesting part is that all the children of the world, French or not, can recognize those differences during the first few months of life. At that point in our development we are able to detect differences that as adults would be impossible for us.

In effect, a baby has a *universal* brain that is able to distinguish phonological contrasts in every language. Over time, each brain develops its own phonological categories and barriers that depend on the specific use of its language. In order to understand that an 'a' pronounced by different people, in varying contexts, at different distances, with head colds and without, corresponds to the same 'a', one has to establish a category of sounds. Doing this means, unfailingly, losing resolution. Those borders for identifying phonemes in the space of sounds are established between six and nine months of life. And they depend, of course, on the language we hear during development. That is the age when our brain stops being universal.

After the early stage in which phonemes are established, it is time for words. Here there is a paradox that, on the face of it, seems hard to resolve. How can babies know which are the words in a language? The problem is not only how to learn the meaning of the thousands of words that make it up. When someone hears a phrase in German for the first time, not only do they not know what each word means but they can't even distinguish them in the sound continuum of the phrase. That is due to the fact that in spoken language there are no pauses that are equal to the space between written words. Thatmeansthatlisteningtosomeonespeakisliketryingtoreadthis.* And if babies don't know which are the words of a language, how can they recognize them in that big tangle?

One solution is talking to babies – as we do when speaking *Motherese* – slowly and with exaggerated enunciation. In *Motherese* there are pauses between words, which facilitates the baby's heroic task of dividing a sentence into the words that make it up.

But this doesn't explain *per se* how eight-month-olds already begin to form a vast repertoire of words, many of which they don't even know how to define. In order to do this, the brain uses a principle similar to the one many sophisticated computers employ to detect patterns, known as statistical learning. The recipe is simple and identifies the frequency of transitions between syllables and function. Since the word *hello* is used frequently, every time the syllable 'hel' is heard, there is a high probability that it will be followed by the syllable 'lo.' Of course, these are just probabilities, since sometimes the word will be *helmet* or *hellraiser*, but a child discovers, through an intense calculation of these transitions, that the syllable 'hel' has a relatively small number of frequent successors. And so, by forming bridges between the most frequent transitions, the child can amalgamate syllables and discover words. This way of learning, obviously not a conscious one, is similar to what *smart*phones use to complete words with the extension they find most probable and feasible; as we know, they don't always get it right.

---

* TheancientGreekswrotelikethiswithoutwordsanditwasallonehieroglyph.

This is how children learn words. It is not a lexical process as if filling a dictionary in which each word is associated with its meaning or an image. To a greater extent, the first approach to words is rhythmic, musical, prosodic. Only later are they tinged with meaning. Marina Nespor, an extraordinary linguist, suggests that one of the difficulties of studying a second language in adulthood is that we no longer use that process. When adults learn a language, they usually do so deliberately and by using their conscious apparatus; they try to acquire words as if memorizing them from a dictionary and not through the musicality of language. Marina maintains that if we were to imitate the natural mechanism of first consolidating the words' music and the regularities in the language's intonation, our process of learning would be much simpler and more effective.

The children of Babel

One of the most passionately debated examples of the collision between biological and cultural predispositions is bilingualism. On one hand, a very common intuitive assumption is: 'Poor child, just learning to talk is difficult, the kid's gonna get all mixed up having to learn two languages.' But the risk of confusion is mitigated by the perception that bilingualism implies a certain cognitive virtuosity.

Bilingualism, actually, offers a concrete example of how some social norms are established without the slightest rational reflection. Society usually considers monolingualism to be the norm, so that the performance of bilinguals is perceived as a deficit or an increment in relation to it. That is not merely convention. Bilingual children have an advantage in the executive functions, but this is never perceived as a deficit in monolinguals' potential development. Curiously, the monolingual norm is not defined by its popularity; in fact, most children in the world grow up being exposed to more than one language. This is especially true in countries with large immigrant populations. In these homes, languages can be combined in all sorts of forms. As a boy, Bernardo Houssay (later awarded the Nobel Prize for Physiology) lived in Buenos Aires, Argentina (where the official language is Spanish) with his Italian grandparents. His parents spoke little of their parents' language, and he and his brothers spoke none. So he believed that people, as they aged, turned into Italians.

Cognitive neuroscientific research has conclusively proven that, going against popular belief, the most important landmarks in language acquisition – the moment of comprehending the first words, the development of sentences, among others – are very similar in monolinguals and bilinguals. One of the few differences is that, during infancy, monolinguals have a bigger vocabulary. However, this effect disappears – and even reverts – when the words a bilingual can use in both languages are added to that vocabulary.

A second popular myth is that one shouldn't mix languages and that each person should speak to a child always in the same language. That is not the case. Some studies in bilingualism are conducted with parents who each speak one language exclusively to their children, which is very typical in border regions, such as where Slovenia meets Italy. In other studies, in bilingual regions such as Quebec or Catalonia, both parents speak both languages. The developmental landmarks in these two situations are identical. And the reason why the babies don't get confused by one person speaking two languages is because, in order to produce the phonemes of each language, they give gesticular indications – the way they move their mouths and face – of which language they are speaking. Let's say that one makes a French or an Italian *facial expression.* These are easy clues for a baby to recognize.

On the other hand, another large group of evidence indicates that bilinguals have a better and faster development of the executive functions; more specifically, in their ability to inhibit and control their attention. Since these faculties are critical in a child's educational and social development, the advantage of bilingualism now seems quite obvious.

In Catalonia, children grow up in a sociolinguistic context in which Spanish and Catalan are often used in the same conversation. As a consequence, Catalan children develop skills to shift rapidly from one language to the other. Will this social learning process extend to task-switching beyond the domain of language?

To answer this question, César Ávila with his colleagues compared brain activity of monolinguals and Catalan bilinguals who switched between non-linguistic tasks. Participants saw a sequence of objects flashing rapidly in the centre of a screen. For a number of trials they were asked to respond with a button if the object was red, and with another button if it was blue. Then, suddenly, participants were asked to forget about colour and respond using the same buttons about the shape of the object (right button for a square and left button for a circle).

As simple as this sounds, when task instructions switch from colour to shape most people respond more slowly and make more errors. This effect is much smaller in Catalonian bilinguals. Ávila also found that the brain networks used by monolinguals and bilinguals to solve this task are very different. It is not that bilinguals are just increasing slightly the amount of activity in one region; it is that the problem in the brain is solved in an altogether different manner.

To switch between tasks, monolinguals use brain regions of the executive system such as the anterior cingulate and some regions in the frontal cortex. Bilinguals instead engage brain regions of the language network, the same regions they engage to switch between Spanish and Catalan in a fluid conversation.

This means that in task-switching, even if the tasks are non-linguistic (in this case switching between colour and shape), bilinguals engage brain networks for language. Which is to say, bilinguals can recycle those brain structures that are highly specialized for language in monolinguals, and use them for cognitive control beyond the domain of language.

Speaking more than one language also changes the brain's anatomy. Bilinguals have a greater density of white matter – bundles of neuronal projections – in the anterior cingulate than monolinguals do. And this effect doesn't pertain only to those who learned more than one language during childhood. It is a characteristic that has been seen also in those who became bilingual later in life, and as such it might be particularly useful in old age, because the integrity of the connections is a decisive element in cognitive reserve. This explains why bilinguals, even when we factor in age, socioeconomic level and other relevant factors, are less prone to developing senile dementias.

To sum up, the study of bilingualism allows us to topple two myths: language development doesn't slow down in bilingual children, and the same person can mix languages with no problem. What's more, the effects of bilingualism may go above and beyond the domain of language, helping develop cognitive control. Bilingualism helps children to be captains of their own thought, pilots of their existence. This ability is decisive in their social inclusion, health and future. So perhaps we should promote bilingualism. Amidst so many less effective and more costly methods of stimulating cognitive development, this is a much simpler, beautiful and enduring way to do so.

A conjecturing machine

Children, from a very young age, have a sophisticated mechanism for seeking out and building knowledge. We were all scientists in our childhood,* and not only out of a desire to explore, to break things apart to see how they work – or used to work – or to pester adults with an infinite number of questions beginning 'Why?' We were also little scientists because of the method we employed to discover the universe.

Science has the virtue of being able to construct theories based on scant, ambiguous data. From the paltry remnants of light from some dead stars, cosmologists were able to build an effective theory on the

---

* Juanjo Sáez, in his lovely book *El Arte*, says: 'I read an interview with Julian Schnabel, artist and film director. He bragged that he started drawing at the age of five. As if he were some sort of precocious genius! What a phoney! We all draw when we're kids and then some of us give it up and others don't.'

origin of the universe. Scientific procedure is especially effective when we know the precise experiment to discriminate between different theories. And kids are naturally gifted at this job.

A game with buttons (push buttons, keys or switches) and functions (lights, noise, movement) is like a small universe. As they play, children make interventions that allow them to reveal mysteries and discover the causal rules of that universe. Playing is discovering. In fact, the intensity of a child's game depends on how much uncertainty the child has with regard to the rules that govern it. And when children don't know how a simple machine works, they usually spontaneously play in the way that is most effective to discover its functioning mechanism. This is very similar to a precise aspect of the scientific method: investigation and methodical exploration in order to discover and clarify causal relationships in the universe.

But children's natural exploration of science goes even further: they construct theories and models according to the most plausible explanation for the data they observe.

There are many examples of this, but the most elegant begins in 1988 with an experiment by Andrew Meltzoff – again – which produced the following scene. An actor enters a room and sits in front of a box with a large plastic button, pushes the button with their head and, as if the box were a slot machine paying out, there is a fanfare with colourful lights and sounds. Afterwards, a one-year-old baby who has been observing the scene is seated, on their mother's lap, in front of the same machine. And then, spontaneously, the young child leans forward and presses the button with their head.

Did they simply imitate the actor or had the one-year-old discovered a causal relationship between the button and the lights? Deciding between these two possibilities would require a new experiment like the one proposed by the Hungarian psychologist György Gergely, fourteen years later. Meltzoff thought that the babies were imitating the actor when they pressed the button with their head. Gergely had another, much bolder and more interesting idea. The babies understand that the adult is intelligent and, because of that, if they didn't push the button with their hand, which would be more natural, it was because pushing it with their head was strictly necessary.

This bold theory suggests that the reasoning of babies turns out to be much more sophisticated, and includes a theory of how things and people work. But how can one detect such reasoning in a child that doesn't yet talk? Gergely solved it in a simple, elegant way. Imagine an analogous situation in everyday life. A person is walking with many bags and opens a door handle with an elbow. We all understand that door handles are not meant to be opened with your elbow and the person did that because there was no other option. What would happen if we replicated this idea in Meltzoff's experiment? The same actor arrives, loaded down with bags, and pushes the button with their head. If the babies are simply imitating the actor, they would do the same. But if, on the other hand, they are capable of thinking logically, they will understand that the actor pushed it with their head because their hands were full and, therefore, all the babies needed to do to get the colourful lights and sounds was to push the button, with any part of their body.

They carried out the experiment. The baby observed the actor, laden with shopping bags, pushing the button with their head. Then the child sits on their mother's lap and pushes the button with their hands. It is the same baby that, upon seeing the actor do the same thing but with their hands free, had pushed the button with their head.

One-year-olds construct theories on how things work based on what they observe. And among those observations is that of perceiving other people's perspectives, working out how much they know, what they can and cannot do. In other words, exploring science.

The good, the bad and the ugly

We began this chapter with the arguments of the empiricists, according to which all logical and abstract reasoning occurs after the acquisition of language. But nevertheless we saw that even newborns form abstract and sophisticated concepts, that they have notions of mathematics, and display some understanding of language. At just a few months old, they already exhibit a sophisticated logical reasoning. Now we will see that young children who do not yet speak have also forged moral notions, perhaps one of the fundamental pillars of human social interaction.

The infants' ideas of what is good, bad, fair, property, theft and punishment – which are already quite well established – cannot be fluently expressed because their control tower (circuits in the prefrontal cortex) is immature. Hence, as occurs with numerical and linguistic concepts, the infants' mental richness of moral notions is masked by their inability to express it.

One of the simplest and most striking scientific experiments to demonstrate babies' moral judgements was done by Karen Wynn in a wooden puppet theatre with three characters: a triangle, a square and a circle. In the experiment, the triangle goes up a hill. Every once in a while it backs up only to later continue to ascend. This gives a vivid impression that the triangle has an intention (climbing to the very top) and is struggling to achieve it. Of course, the triangle doesn't have real desires or intentions, but we spontaneously assign it beliefs and create narrative explanations of what we observe.

A square shows up in the middle of this scene and bumps into the triangle on purpose, sending it down the hill. Seen with the eyes of an adult, the square is clearly despicable. As the scene is replayed, the circumstances change. While the triangle is going up, a circle appears and pushes it upwards. To us the circle becomes noble, helpful and gentlemanly.

This conception of good circles and bad squares needs a narrative – which comes automatically and inevitably to adults – that, on the one hand, assigns intentions to each object and, on the other, morally judges each entity based on those intentions.

As humans, we assign intentions not only to other people but also to plants ('sunflowers seek out the sun'), abstract social constructions ('history will absolve me' or 'the market punishes investors'), theological entities ('God willing') and machines ('damn washing machine'). This ability to theorize, to turn data into stories, is the seed of all fiction. That is why we can cry in front of a television set – it is strange to cry because something happens to some tiny pixels on a screen – or destroy blocks on an iPad as if we were in a trench on the Western Front during the First World War.

In Wynn's puppet show there are only triangles, circles and squares, but we *see* them as someone struggling, a bad guy who hinders progress, and a do-gooder who helps. Which is to say that, as adults, we have an automatic tendency to assign moral values. Do six-month-olds have that same abstract thought process? Would babies be able spontaneously to form moral conjectures? We can't know by asking because they don't yet talk, but we can infer this narrative by observing their preferences. The constant secret of science consists, precisely, in finding a way of bridging what we want to know – in this case, whether babies form moral concepts – with what we can measure (which objects the babies choose).

After watching one object helping the circle climb the hill and another bumping it down, infants were encouraged to reach for one of them. Twenty-six of twenty-eight (twelve out of twelve six-month-olds) chose the helper. Then, the video recordings of the infants watching the scenes of the helper and the hinderer were shown to an experimentalist. And, relying on their facial gestures and expressions alone, she could predict almost perfectly whether the infant had just seen the helper or the hinderer.

Six-month-old infants, before crawling, walking or talking, when they are barely discovering how to sit up and eat with a spoon, are already able to infer intentions, desires, kindness and evil, as can be deduced from examining their choices and gestures.

He who robs a thief . . .

The construction of morality is, of course, much more sophisticated. We cannot judge a person to be good or bad just by knowing they did something helpful. For example, helping a thief is usually considered ignoble. Would the babies prefer someone who helps a thief to someone who thwarts one? We are now in the murky waters that are the origins of morality and law. But even in this sea of confusion, babies between nine months and a year of age already have an established opinion.

The experiment that proves it goes like this. Babies see a hand puppet trying to lift the top off a box in order to pull out a toy. Then a helpful puppet shows up and helps it open the lid and get the toy. But in another scene an anti-social puppet jumps maliciously on to the box, slamming it shut and keeping the first puppet from getting at the toy. When choosing between the two puppets, the babies prefer the helper. But here Wynn was going for something much more interesting: identifying what the babies think about stealing from an evildoer, long before they know those words.

To do this she designed a third act for the puppet theatre, and the helper puppet now loses a ball. In some cases, in this garden of forking paths, a new character appears on the scene and returns the ball. At other times, another character comes in, steals it and runs away. The babies prefer the character that returns the ball.

But the most interesting and mysterious part happens when these scenes feature the antisocial puppet that jumped maliciously on the box. In this case, the babies change their preference. They sympathize with the one who steals the ball and runs away. For nine-month-olds, the one who gives the bad guy his comeuppance is more lovable than the one who helps him, at least in that world of puppets, boxes and balls.*

Preverbal babies, still unable to coordinate their hands in order to grab an object, do something much more sophisticated than judging others by their actions. They take into account the contexts and the history, which turns out to give them a pretty sophisticated notion of justice. That's how incredibly disproportionate cognitive faculties are during the early development of a human being.

The colour of a jersey, strawberry or chocolate

We adults are not unbiased when we judge others. Not only do we keep in mind their previous history and the context of their actions (which we should), but we also have very different opinions of the person committing the actions, or being the victim of them, if they look like us or not (which we shouldn't).

Throughout all cultures, we tend to form more friendships and have more empathy with those who look like us. On the other hand, we usually judge more harshly and show more indifference to the suffering of those who are different. History is filled with instances in which human groups have massively supported or, in the best-case scenario, rejected violence directed at individuals who were not like them.

This even manifests itself in formal justice proceedings. Some judges serve sentences displaying a racial bias, most probably without being aware that race is influencing their judgement. In the United States, African American males have been incarcerated at about six times the rate of white males. Is this difference in the incarceration rate a result (at least in part) of the judges having different sentencing

---

* . . . in which we live.

practices? This seemingly simple and direct question turns out to be hard to answer because it is difficult to separate this psychological factor from possible racial differences in case characteristics. To overcome this problem Sendhil Mullainathan, Professor of Economics at Harvard University, found an ingenious solution, exploiting the fact that in the United States cases are randomly assigned to judges. Hence, on average, the type of case and the nature of defendants are the same for all judges. A racial difference in sentencing could potentially be explained by case characteristics or by a difference in the quality of the assigned attorneys (which is not random). But if this were all, then this difference should be the same for all judges. Instead, Mullainathan found a huge disparity – of almost 20 per cent – between judges in the racial gap in sentencing. While this may be the most convincing demonstration that race matters in the courtroom, the method is partly limited since it cannot tell whether the variability between judges' results is due to some of them discriminating against African Americans, or some judges discriminating against whites, or a mixture of both.

Physical appearance also affects whether someone is likely to be hired in a job interview. Since the early seventies, several studies have shown that attractive applicants are typically judged to have a more appropriate personality for a job, and to perform better than their less attractive counterparts. Of course, this was not just a matter of comment. Applicants who were judged to be more attractive were also more likely to be hired. As we will see in Chapter 5, we all tend to make retrospective explanations that serve to justify our choices. Hence the most likely timeline for this line of argument is like this: first the interviewer decides to hire the applicant (among other things based on his or her beauty) and only then generates ad hoc a long list of attributes (he or she was more capable, more suited for the job, more reliable . . .) that serve to justify the choice which indeed had nothing to do with these considerations.

The similarities that generate these predispositions can be

based on physical appearance, but also on religious, cultural, ethnic, political or even sports-related questions. This last example, because it is presumed to be more harmless – although, as we know, even sporting differences can have dramatic consequences – is easier to assimilate and recognize. Someone forms part of a consortium, a club, a country, a continent. That person suffers and celebrates collectively with that consortium. Pleasure and pain are synchronized between thousands of people whose only similarity is belonging to a tribe (sharing a jersey, a neighbourhood or a history) that unites them. But there is something more: pleasure at the suffering of other tribes. Brazil celebrates Argentina's defeats, and Argentina celebrates Brazil's. A fan of Liverpool cheers for the goal scored against Manchester United. When rooting for our favourite sports teams, we often feel less inhibited about expressing *Schadenfreude*, our pleasure at the suffering of those unlike us.

What are the origins of this? One possibility is that it has ancestral evolutionary roots, that the drive collectively to defend what belonged to one's tribe was advantageous at some point in human history and, as a result, adaptive. This is merely conjecture but it has a precise, observable footprint that can be traced. If *Schadenfreude* is a constituent aspect of our brains (the product of a slow learning process within evolutionary history), it should manifest itself early in our lives, long before we establish our political, sports or religious affiliations. And that is exactly how it happens.

Wynn performed an experiment to ask whether infants prefer those who help or harm dissimilar persons. This experiment was also carried out in a puppet theatre. Before entering the theatre, a baby between nine and fourteen months old, seated comfortably on their mother's lap, chose between crackers or green beans. Apparently, food choices reveal tendencies and strong allegiances.

Then two puppets came in, successively and with a considerable amount of time between the two entrances. One puppet demonstrates an affinity with the baby and says that it loves the food the child has chosen. Then they leave and, just as before, there is another scene where the puppet with similar taste is playing with a ball, drops it and has to deal with two different puppets: one who helps and the other who steals the ball. Then babies are asked to pick up one of the two puppets and they show a clear preference for the helper. One who helps someone similar to us is good. But when the puppet who loses the ball is the one who had chosen the other food, the babies more often choose the ball robber. As with the thief, it is gastronomic *Schaden-freude*: the babies sympathize with the puppet that hassles the one with different taste preferences.

Moral predispositions leave robust, and sometimes unexpected, traces. The human tendency to divide the social world into groups, to prefer our own group and go against others, is inherited, in part, from predispositions that are expressed very early in life. One example that has been particularly well studied is language and accent. Young children look more at a person who has a similar accent and speaks their mother tongue (another reason to advocate bilingualism). Over time, this bias in our gazes disappears but it transforms into other manifestations. At two years old, children are more predisposed to accept toys from those who speak their native language. Later, at school age, this effect becomes more explicit in the friends they choose. As adults, we are already familiar with the cultural, emotional, social and political segregations that emerge simply based on speaking different languages in neighbouring regions. But this is not only an aspect of language. In general, throughout their development, children choose to relate to the same type of individuals they would have preferentially directed their gaze at in early childhood.

As happens with language, these predispositions develop, transform and reconfigure with experience. Of course, there is nothing within us that is exclusively innate; to a certain extent, everything takes shape on the basis of our cultural and social experience. This book's premise is that revealing and understanding these predispositions can be a tool for changing them.

Émile and Minerva's owl

In *Émile, or Concerning Education*, Jean-Jacques Rousseau sketches out how an ideal citizen should be educated. The education of Émile would today be considered somewhat exotic. During his entire childhood there is no talk of morality, civic values, politics or religion. He never hears the arguments we parents of today so often go on about, like how we have to share, be considerate of others, among so many other outlines of arguments for fairness. No. Émile's education is far more similar to the one Mr Miyagi gives Daniel LaRusso in *The Karate Kid*, pure praxis and no words.

So, through experience, Émile learns the notion of property at twelve years old, at the height of his enthusiasm for his vegetable garden. One day he shows up with watering-pot in hand and finds his garden plot destroyed.

> But oh, what a sight! What a misfortune! [. . .] What has become of my labour, the sweet reward of all my care and toil? Who has robbed me of my own? Who has taken my beans away from me? The little heart swells with the bitterness of its first feeling of injustice.*

Émile's tutor, who destroyed his garden on purpose, conspires with the gardener, asking him to take responsibility for the damage and gives him a reason to justify it. Thus the gardener accuses Émile of having ruined the melons that he'd planted earlier in the same plot. Émile finds himself embroiled in a conflict between two legal principles: his conviction that the beans belong to him because he toiled

---

* Translation by Eleanor Worthington.

to produce them and the gardener's prior right as legitimate owner of the land.

The tutor never explains these ideas to Émile, but Rousseau maintains that this is the best possible introduction to the concept of ownership and responsibility. As Émile meditates on this painful situation of loss and the discovery of the consequences of his actions on others, he understands the need for mutual respect in order to avoid conflicts like the one has just suffered. Only after having lived through this experience is he prepared to reflect on contracts and exchanges.

The story of Émile has a clear moral: not to saturate our children with words that have no meaning for them. First they have to learn what they mean through concrete experience. Despite this being a recurrent intuition in human thought, repeated in various landmark texts of the history of philosophy and education,* today hardly anyone follows that recommendation. In fact, almost all parents express an endless enumeration of principles through discourse that we contradict with our actions, such as on the use of telephones, what we should eat, what we should share, how we should say *thank you*, *sorry* and *please*, not be insulting, etc.

I have the impression that the entire human condition can be expressed with a piñata. If a Martian arrived and saw the highly complex situation that suddenly arises when the papier mâché breaks and

---

* Rousseau's ideas in *Émile* drew upon the views of Plato. Education must begin with music, gymnastics and other practical matters that train the virtues of a good citizen, in his *Republic*. Only after having walked that long road is one ready to understand the *episteme*, true knowledge. Hegel also advocated educating first through action and only afterwards through discourse. Knowledge is acquired through experiences lived during the day, and theory awakes only as dusk falls, like Minerva's owl. This idea has made itself felt in authors such as Paul Tough and Ken Robinson, who suggest that education should be focused less on knowledge (mathematics, language, history, geography) and more on practice that promotes virtues such as motivation, control and creativity.

the rain of sweets falls out, it would understand all of our yearnings, vices, compulsions and repressions. Our euphoria and our melancholy. It would see the children scrambling to gather up the sweets until their hands can't hold any more; the one who hits another to gain a time advantage over a limited resource; the father who lectures another kid to share their excessive haul; the overwhelmed youngster crying in a corner; the exchanges on the official market and the black market, and the societies of parents who form like micro-governments to avoid what Garrett Hardin called the *tragedy of the commons.*

I, me, mine and other permutations by George

Long before becoming great jurists, philosophers, or noted economists, children – including the children that Aristotle, Plato and Piaget once were – already had intuitions about property and ownership. In fact, children use the pronouns *my* and *mine* before using *I* or their own names. This language progression reflects an extraordinary fact: the idea of ownership precedes the idea of identity, not the other way around.

In early battles over property the principles of law are also rehearsed. The youngest children claim ownership of something based on the argument of their own desires: 'It's mine because I want it.'* Later, around two years of age, they begin to argue with an acknowledgement of others' rights to claim the same property for themselves. Understanding others' ownership is a way of discovering that there are other individuals. The first arguments outlined by children are usually: 'I had it first'; 'They gave it to me.' This intuition that the first person to touch something wins indefinite rights to its usage does not disappear in adulthood. Heated discussions over a parking spot, a seat on a bus, or the ownership of an island by the first country to plant

---

* This is the cry of children under eighteen months old when a toy is taken from them. This is the expression of the only argument that underpins their ownership: their desire for it.

its flag there are private and institutional examples of these heuristics. Perhaps because of that, it is unsurprising that large social conflicts, like in the Middle East, are perpetuated by very similar arguments to those deployed in a dispute between two-year-olds: 'I got here first'; 'They gave it to me.'

Transactions in the playground, or the origin of commerce and theft

On the local 5-a-side football pitch, the owner of the ball is, to a certain extent, also the owner of the game. It gives them privileges like deciding the teams, and declaring when the game ends. These advantages can also be used to negotiate. The philosopher Gustavo Faigenbaum, in Entre Ríos, Argentina, and the psychologist Philippe Rochat, in Atlanta, in the USA, set out to understand this world: basically, how the concept of owning and sharing is established in children, among intuitions, praxis and rules. Thus they invented the *sociology of the playground*. Faigenbaum and Rochat, in their voyage to the land of childhood,* researched swapping, gifts and other transactions that took place in a primary school playground. Studying the exchange of little figurines, they found that even in the supposedly naïve world of the playground, the economy is formalized. As children grow up, lending and the assignment of vague, future values give way to more precise exchanges, the notion of money, usefulness and the prices of things.

As in the adult world, not all transactions in the country of childhood are licit. There are thefts, scams and betrayals. Rousseau's conjecture is that the rules of citizenship are learned in discord. And

---

* As an economist, Paul Webley, says: 'Childhood is another country and they do things differently there. What is required to interpret this culture are local informants. This suggests that child collaborators are vital, and that without them we many find ourselves left outside the gates of the playground, staring in.'

it is the playground, which is more innocuous than real life, that becomes the breeding ground in which to play at the game of law.

The contrasting observations of Wynn and her colleagues suggest that very young children should already be able to sketch out moral reasoning. On the other hand, the work of Piaget, who is an heir to Rousseau's tradition, indicates that moral reasoning only begins at around six or seven years old. Gustavo Faigenbaum and I set out to reconcile these different great thinkers in the history of psychology. And, along the way, to understand how children become citizens.

> We showed to a group of children between four and eight years of age a video with three characters: one had chocolates, the other asked to borrow them and the third stole them. Then we asked a series of questions to measure varying degrees of depth of moral comprehension; if they preferred to be friends with the one who stole or the one who borrowed* (and why), and what the thief had to do to make things right with the victim. In this way we were able to investigate the notion of justice in playground transactions.

Our hypothesis was that the preference for the borrower over the thief, an implicit manifestation of moral preferences – as in Wynn's experiments – should already be established even for the younger kids. And, to the contrary, the justification of these options and the understanding of what had to be done to compensate for the damage caused – as in Piaget's experiments – should develop at a later stage. That is exactly what we proved. In the room with the four-year-olds, the children preferred to play with the borrower rather than with the thief. We also discovered that they preferred to play with someone

---

* Of course, we didn't use those words in the experiment, in order to avoid suggesting preferences through language. Each character had a name and the gender of the borrower or thief changed for different children, so that we could ensure there was no bias in our research.

who stole under extenuating circumstances than with aggravating ones.

But our most interesting finding was this: when we asked four-year-old children why they chose the borrower over the thief or the one who robbed in extenuating circumstances over the one who did so in aggravating ones, they gave responses like 'Because he's blond' or 'Because I want her to be my friend.' Their moral criteria seemed completely blind to causes and reasons.

Here we find again an idea which has appeared several times in this chapter. Children have very early (often innate) intuitions – what the developmental psychologists Liz Spelke and Susan Carey refer to as core knowledge. These intuitions are revealed in very specific experimental circumstances, in which children direct their gaze or are asked to choose between two alternatives. But core knowledge is not accessible on demand in most real-life situations where it may be needed. This is because at a younger age core knowledge cannot be accessed explicitly and represented in words or concrete symbols.

Specifically, in the domain of morality, our results show that children have from a very young age intuitions about ownership which allow them to understand whether a transaction is licit or not. They understand the notion of theft, and they even comprehend subtle considerations which extenuate or aggravate it. These intuitions serve as a scaffold to forge, later in development, a formal and explicit understanding of justice.

But every experiment comes with its own surprises, revealing unexpected aspects of reality. This one was no exception. Gustavo and I came up with the experiment to study *the price of theft*. Our intuition was that the children would respond that the chocolate thief should give back the two they stole plus a few more as compensation for the damages. But that didn't happen. The vast majority of the children felt that the thief had to return exactly the two chocolates that had been stolen. What's more, the older the kids, the higher the fraction of those who advocated an exact restitution. Our hypothesis was mistaken. Children are much more morally dignified than we had imagined.

They understood that the thieves had done wrong, that they would have to make up for it by returning what they'd stolen along with an apology. But the moral cost of the theft could not be resolved in kind, with the stolen merchandise. In the children's justice, there was no reparation that absolved the crime.

If we think about the children's transactions as a toy model of international law, this result, in hindsight, is extraordinary. An implicit, though not always respected, norm of international conflict resolution is that there should be no escalation in reprisal. And the reason is simple. If someone steals two and, in order to settle a peace, the victim demands four, the exponential growth of reprisals would be harmful for everyone. Children seem to understand that even in war there ought to be rules.

Jacques, innatism, genes, biology, culture and an image

Jacques Mehler is one of many Argentinian political and intellectual exiles. He studied with Noam Chomsky at the Massachusetts Institute of Technology (MIT) at the heart of the cognitive revolution. From there he went to Oxford and then France, where he was the founder of the extraordinary school of cognitive science in Paris. He was exiled not just as a person, but as a thinker. He was accused of being a reactionary for claiming that human thought had a biological foundation. It was the oft-mentioned divorce between human sciences and exact sciences, which in psychology was particularly marked. I like to think of this book as an ode to and an acknowledgment of Jacques's career. A space of freedom earned by an effort that he began, swimming against the tide. An exercise in dialogue.

In the epic task of understanding human thought, the division between biology, psychology and neuroscience is a mere declaration of caste. Nature doesn't care a fig for such artificial barriers between types of knowledge. As a result, throughout this chapter, I have interspersed biological arguments, such as the development of the frontal cortex, with cognitive arguments, such as the early development of moral notions. In other examples, like that of bilingualism and attention, we've delved into how those arguments combine.

Our brains today are practically identical to those of at least 60,000 years ago, when modern man migrated from Africa and culture was completely different. This shows that individuals' paths and potential for expression are forged within their social niches. One of the arguments of this book is that it is also virtually impossible to understand human behaviour without taking into consideration the traits of the organ that comprises it: the brain. The way in which social knowledge and biological knowledge interact and complement each other depends, obviously, on each case and its circumstances. There are some cases in which biological constitution is decisive. And others are determined primarily by culture and the social fabric. It is not very different from what happens with the rest of the body. Physiologists and coaches know that physical fitness can change enormously during our life while, on the other hand, our running speed, for example, doesn't have such a wide range of variation.

The biological and the cultural are always intrinsically related. And not in a linear manner. In fact, a completely unfounded intuition is that biology precedes behaviour, that there is an innate biological predisposition that can later follow, through the effect of culture, different trajectories. That is not true; the social fabric affects the very biology of the brain. This is clear in a dramatic example observed in the brains of two three-year-old children. One is raised with affection in a normal environment while the other lacks emotional, educational and social stability. The brain of the latter is not only abnormally small but its ventricles, the cavities through which cerebrospinal fluid flows, have an abnormal size as well.

So different social experiences result in completely distinct brains. A caress, a word, an image – every life experience leaves a trace in the brain. These traces modify the brain and, with it, one's way of responding to things, one's predisposition to relating to someone, one's desires, wishes and dreams. In other words, the social context changes the brain, and this in turn defines who we are as social beings.

A second unfounded intuition is thinking that because something is biological it is unchangeable. Again, this is simply not true. For instance, the predisposition to music depends on the biological constitution of the auditory cortex. This is a causal relation between an organ and a cultural expression. However, this connection does not imply developmental determinism. The auditory cortex is not static, anyone can change it just by practising and exercising.

Thus the social and the biological are intrinsically related in a network of networks. This categorical division is not a property of nature, but rather of our obtuse way of understanding it.

CHAPTER TWO

The fuzzy borders of identity

What defines our choices and allows us to trust
other people and our own decisions?

Our choices define us. We choose to take risks or live conservatively,
to lie when it seems convenient or to make the truth a priority, no
matter what the cost. We choose to save up for a distant future or live
in the moment. The vast sum of our actions comprises the outline
of our identities. As José Saramago put it in his novel *All the Names*:
'We don't actually make decisions, the decisions make us.' Or, in a
more contemporary version, when Albus Dumbledore lectures Harry
Potter: 'It is our choices, Harry, that show what we truly are, far more
than our abilities.'

Almost all decisions are mundane, because the overwhelming
majority of our lives are spent in the day-to-day. Deciding whether
we'll visit a friend after work, if we should take the bus or the Underground; choosing between chips or a salad. Imperceptibly, we compare
the universe of possible options on a mental scale, and after thinking
it over we finally choose (chips, of course). When choosing between
these alternatives, we activate the brain circuits that make up our
mental decision-making machine.

Our decisions are almost always made based on incomplete
information and imprecise data. When a parent chooses what school
to send their child to, or a Minister of Economics decides to change

the tax policy, or a football player opts to shoot at goal instead of passing to a teammate in the penalty area – in each and every one of these occasions it is only possible to sketch an approximate idea of the impending consequences of our decisions. Making decisions is a bit like predicting the future, and as such is inevitably imprecise. *Eppur si muove*. The machine works. That is what's most extraordinary.

Churchill, Turing and his labryinth

On 14 November 1940, some 500 Luftwaffe planes flew, almost unchallenged, to Britain and bombed the industrial city of Coventry for seven hours. Many years after the war had ended, Captain Frederick William Winterbotham revealed that Winston Churchill* could have avoided the bombing and the destruction of the city if he had decided to use a secret weapon discovered by the young British mathematician Alan Turing.

Turing had achieved a scientific feat that gave the Allies a strategic advantage that could decide the outcome of the Second World War. He had created an algorithm capable of deciphering Enigma, the sophisticated mechanical system made of circular pieces – like a combination lock – that allowed the Nazis to encode their military messages. Winterbotham explained that, with Enigma decoded, the secret service men had received the coordinates for the bombing of Coventry with enough warning to take preventive measures. Then, in the hours leading up to the bombing, Churchill had to decide between two options: one emotional and immediate – avoiding the horror of a civilian massacre – and the other rational and calculated – sacrificing

---

* In the book in which Churchill gives his views on the Second World War – and which earned him the Nobel Prize for Literature – he doesn't mention this now controversial story. Actually, Churchill doesn't talk about any intelligence operation, after realizing the mistake he had made by revealing information in his book on the First World War, which ended up being useful for the Axis during the Second World War.

Coventry, not revealing their discovery to the Nazis, and holding on to that card in order to use it in the future. Churchill decided, at a cost of 500 civilian lives, to keep Britain's strategic advantage over his German enemies a secret.

Turing's algorithm evaluated in unison all the configurations – each one corresponding to a possible code – and, according to its capacity to predict a series of likely messages, updated each configuration's probability. This procedure continued until the likelihood of one of the configurations reached a sufficiently high level. The discovery, in addition to precipitating the Allied victory, opened up a new window for science. Half a century after the war's end it was discovered that the algorithm that Turing had come up with to decode Enigma was the same one that the human brain uses to make decisions. The great English mathematician, who was one of the founders of computation and artificial intelligence, created – in the urgency of wartime – the first, and still the most effective, model for understanding what happens in our brains when we make a decision.

Turing's brain

As in the procedure sketched out by Turing, the cerebral mechanism for making decisions is built on an extremely simple principle: the brain elaborates a landscape of options and starts a winner-take-all race between them.

The brain converts the information it has gathered from the senses into votes for one option or the other. The votes pile up in the form of ionic currents accumulated in a neuron until they reach a threshold where the brain deems there is sufficient evidence. These circuits that coordinate decision-making in the brain were discovered by a group of researchers headed by William Newsome and Michael Shadlen. Their challenge was to design an experiment simple enough to be able to isolate each element of the decision and, at the same time, sophisticated enough to represent decision-making in real life.

This is how the experiment works: a cloud of dots moves on a screen. Many of the dots move in a chaotic, disorganized way. Others move coherently, in a single direction. A player (an adult, a child, a monkey and, sometimes, a computer) decides which way that cloud of dots is moving. It is the electronic version of a sailor lifting a finger to decide, in the midst of choppy waters, which way the wind is blowing. Naturally, the game becomes easier when more dots are moving in the same direction.

Monkeys played this game thousands of times, while the researchers recorded their neuronal activity as reflected by the electrical currents produced in their brains. After studying this exercise for many years, and in many variations, they revealed the three principles of Turing's algorithm for decision-making:

(1) A group of neurons in the visual cortex receives information from the retina. The neuron's current reflects the quantity and direction of movement in each moment, but does not accumulate a history of these observations.

(2) The sensory neurons are connected to other neurons in the parietal cortex, which amass this information over time. So the neuronal circuits of the parietal cortex codify how the predisposition towards each possible action changes over time during the course of making the decision.

(3) As information favouring one option accumulates, the parietal cortex that codifies this option increases its electrical activity. When the activity reaches a certain threshold, a circuit of neurons in structures deep in the brain – known as basal ganglia – set off the corresponding action and restart the process to make way for the next decision.

The best way to prove that the brain decides through a race in the parietal cortex is by showing that a monkey's response can be conditioned by injecting a current into the neurons that codify evidence in favour of a certain option. Shalden and Newsome did that experiment. While one monkey was watching a cloud of dots that moved completely randomly, they used an electrode to inject an electrical current into the parietal neurons that codify movement to the right. And, despite the senses indicating that movement was tied in either direction, the monkeys always responded that they were moving to the right. This is like emulating electoral fraud, manually inserting certain votes into the ballot box.

Additionally, this series of experiments allowed for the identification of three fundamental traits of the decision-making process. What relationship is there between the clarity of the evidence and the time we take to make a decision? How are options biased by prejudices or prior knowledge? When is there enough evidence in favour of one option to call the race? The answers to these three questions are interrelated. The more incomplete the information is, the slower the accumulation of evidence will be. In the moving-dot experiment, when almost all the dots move at random, the *ramp* of activation in the neurons in the parietal cortex that amass the evidence is not very steep. And if the threshold of evidence needed remains the same, it will take more time to cross it; which is to say, to reach the same degree of reliability. The decision cooks over a slow flame, but eventually it will reach the same temperature.

And how is the threshold established? Or, to put it another way, how does the brain determine when enough is enough? This depends on a calculation that the brain makes in a stunningly precise way, by pondering the cost of making a mistake and the time available for the decision-making.

The brain determines that threshold in order to optimize the gains from a decision. To do so it combines neuronal circuits that codify:

(1) The value of the action.
(2) The cost of time invested.

(3) The quality of the sensory information.
(4) An endogenous urgency to respond, something that we recognize as anxiety or impatience to decide.

If, in the random-dots game, mistakes are punished severely, the players (humans or monkeys) raise the threshold, taking more time to decide and accumulating more evidence. If, on the other hand, mistakes don't count, then the players lower that same threshold, adopting again the best strategy, which here is to respond as quickly as possible. The most notable aspect of this adaptive adjustment is that in most cases it is not conscious, and often far more optimal than we would imagine.

Consider, for example, a driver stopping at a traffic light. The driver's brain is making a great number of estimations: the probability that the light may turn amber or red, the distance to the crossing, the speed of the car, the effectiveness of the brakes, the traffic etc. Not only this: the driver´s brain is also pondering the urgency, the consequences of an accident . . . In the vast majority of cases (except when something goes wrong and the monitoring system of the brain takes control) these considerations are not explicit. We are not aware of all these calculations. Yet our brains do make this sophisticated calculus, which results in a decision of when and how hard we will hit the brake pedal. This specific example reveals a general principle: decision-makers know much more than they believe they do.

In contrast with this, in some conscious deliberations (which are the only ones we do remember at the end of the day) the brain often sets a very inefficient threshold to reach a decision. We all remember having slept too long on some matters which did not require that much deliberation. For example, most of us recall deliberating ad infinitum in a restaurant between two choices even if deep inside we know we would greatly enjoy either of those two options.

Turing in the supermarket

Even though in the laboratory we study simple decisions, what we are ultimately more interested in revealing is how the brain makes everyday decisions: the driver who decides whether or not to jump an amber light; the judge who condemns or exonerates a defendant; the voter who casts a ballot for one candidate or another; the shopper who takes advantage of or falls victim to a special deal. The conjecture is that all of these decisions, despite belonging to different realms and having their own idiosyncrasies, are the result of the same decision-making mechanism.

One of the main principles of this procedure, which is at the heart of Turing's design, consists in how one realizes when it is time to stop gathering evidence. The problem is reflected in the paradox described by a medieval philosopher, Jean Buridan: a donkey hesitates end-lessly between two identical piles of hay and, as a result, ends up dying of hunger. In fact, the paradox presents a problem for Turing's pure model. If the number of votes in favour of each alternative is identical, the cerebral race is stuck in a tie. The brain has a way of avoiding the tie: when it considers that sufficient time has passed, it invents neuro-nal activity that it randomly distributes among the circuits that codify each option. Since this current is random, one of the options ends up having more votes and, as such, wins the race. It's as if the brain tossed a coin and let fate break the tie. How much time is reasonable for mak-ing a decision depends on internal states of the brain – for example, if we are more or less anxious – and on external factors that affect how the brain counts the time.

One of the ways that the brain estimates time is simply by count-ing pulses: steps, heartbeats, breaths, the swinging of a pendulum or music's tempo. For example, when we exercise, we mentally estimate a minute faster than when we are at rest, because each heartbeat – and therefore each pulse of our inner clock – is quicker. The same happens with tempo in music. The clock accelerates with the rhythm and, thus, time passes more rapidly. Do these changes in our internal clock make us decide more quickly and lower our decision threshold?

Indeed, music has much more direct consequences for our decisions than we recognize. We drive, shop and walk differently depending on the music we are listening to at the time. As the musical tempo rises, our decision-making threshold lowers and as a result risk increases in almost every decision. Drivers change lanes more frequently, go through more amber lights, overtake and exceed the speed limit more while driving as the speed of the music they are listening to increases. Musical tempo also dictates the amount of time we are willing to wait patiently in a waiting room or the number of products we tend to buy in a supermarket. Many supermarket managers know that the piped-in music is a key to sales and use that to their advantage, with no need to be familiar with Turing's work. That's how predictable our decision-making machine is, yet we are almost completely unaware of its workings.

Another key factor that affects the decision-making machine is determining where the race begins. When there is a bias towards one of the alternatives, the neurons that accumulate information in its favour start with an initial electrical charge, which is similar to giving them a head start in the race. In some cases, biases can have a fundamental influence; for example, in the decision to donate organs.

Demographic studies of organ donation group different countries into two classes: those in which almost all the inhabitants agree to donate organs, and another in which almost no one does. It doesn't take a master statistician to understand that what's striking is the absence of intermediate classes. The reason turns out to be extremely simple: what ends up determining whether a person chooses to donate organs is the wording on the form. In the countries where the form says: 'If you wish to donate organs, sign here', no one does. On the other hand, in countries where it says: 'If you do NOT wish to donate organs, sign here', almost everyone donates. The explanation for both phenomena comes from an almost universal trait that has nothing to do with religion or life and death but rather just that no one fills out the form completely.

When we are offered a wide variety of options, they don't all start running from the same point; those that are given by default begin with an advantage. If, in addition, the problem is one that is hard to resolve, meaning that evidence in favour of any of the options is scarce, the one that started out with the advantage wins. This is a very clear example of how governments can guarantee freedom of choice but, at the same time, bias – and, in practice, dictate – what we decide. But this also reveals a characteristic of human beings, be they Dutch, Mexican, Catholic, Protestant or Muslim: our decision-making mechanism collapses when faced with difficult situations. Then we merely accept what we are offered, by default.

The tell-tale heart

Until now we've talked about decision-making processes as if they were all of one class, governed by the same principles and carried out in the brain by similar circuits. However, we all perceive that the decisions we make belong to at least two qualitatively distinct types; some are rational and we can put forward the arguments behind them. Others are hunches, inexplicable decisions that feel as if they are dictated by our bodies. But are there really two different ways of deciding? Is it better to choose something based on our intuitions, or to carefully and rationally deliberate each decision?

In general we associate rationality with science, while the nature of our emotions seems mysterious, esoteric and essentially inexplicable. We will topple this myth with a simple experiment.

Two neuroscientists, Lionel Naccache and Stanislas Dehaene – my mentor in Paris – did an experiment in which they flashed numbers on screens so fleetingly that the participants believed they'd seen nothing. This type of presentation, which doesn't activate consciousness, is called subliminal. Then they ask the participants to say if the number is higher or lower than five and, much to their own surprise, they answer correctly in most cases. The person making the decision perceives it as a hunch, but from the experimenter's perspective it is clear that the decision was induced unconsciously with a mechanism very similar to that of conscious decision-making.

Which is to say that, in the brain, hunches aren't so different from rational decisions. But the previous example doesn't capture all the richness of the physiology of unconscious decisions. In this case, popular expressions such as 'trust your heart' or 'go with your gut feelings' turn out to be quite accurate and shed light on how intuitions are forged.

All it takes to understand this is putting a pencil between your teeth, lengthwise. Inevitably, your lips will rise in an imitation of a smile. This is obviously a mechanical effect, not a reflection of an emotion. But that doesn't matter, it still gives a certain sense of wellbeing. The mere gesture of the smile is enough. A film scene will seem more entertaining to us if we watch it with a pencil held in our mouth that way than if we hold it between our lips, as if scowling. So, deciding whether something is fun or boring does not only originate in an evaluation of the external world, but in visceral reactions produced in our internal worlds. Crying, sweating, trembling, increasing heart rate or secreting adrenaline are not merely reactions by the body to communicate an emotion. Instead, the brain reads and identifies these bodily variables to encode and produce feelings and emotions.

That corporeal states can affect our decision-making process is a physiological and scientific demonstration of what we perceive as a hunch. When making a decision unconsciously, the cerebral cortex evaluates different alternatives and, in doing so, estimates the possible risks and benefits of each option. The result of this computation is expressed in corporeal states through which the brain can recognize risk, danger or pleasure. The body becomes a reflection and a resonating chamber of the external world.

The body in the casino and at the chessboard

The key experiment showing how decisions are based on hunches was done with two decks of cards.

As in so many board games, this experiment employs ingredients from real life decision-making: winnings, losses, uncertainty and risk. The game is simple but unpredictable. In each turn, the player merely chooses which deck to pick a card from. The number of the card chosen indicates the coins that the player wins (or loses if it's negative). Since the cards are turned face down, the player has to evaluate, over the course of the entire experiment, which of the two decks is more profitable.

This is like someone in a casino who has to choose between two one-armed bandits just by observing how many times and how much each one pays out over a period of time. But, unlike in the casino, this game thought up by a neurobiologist, Antonio Damasio, is not purely random: there is one deck that on average pays out more than the other. If this rule is discovered, then the next step is simple: always choose the deck that pays more. Lo and behold, an infallible system.

The difficulty lies in the fact that the player has to discover this rule through pondering a long history of payouts amid large fluctuations. After much practice, almost everyone discovers the rule, is able to explain it and, naturally, starts to choose cards from the correct deck every time. But the real finding happens along the way to this discovery, among intuitions and hunches. Even before being able to articulate the rule, the players start to play well and more frequently choose cards from the correct deck. In this phase, despite playing much better than when they were choosing randomly, the players cannot explain why they opt for the correct deck (the one that pays out more in the long term). Sometimes they don't even know they are choosing one deck more than the other. But unequivocal signs show up in their bodies. In this part of the experiment, when players are about to choose the incorrect deck, their skin conductance increases, indicating a rise in sweating, which is in turn a reflection of an emotional state. Which is to say that the players cannot explain that one of the decks gives better results than the other, but their bodies already know it.

My colleague María Julia Leone, a neuroscientist and international chess master, and I carried out this experiment on the chessboard, following the Borgesian concept of chess as a metaphor for life. Two masters face off. They have thirty minutes to make a series of decisions that will organize their armies. On the board, it is a battle to the death and emotions are running high. During the game we trace the players' heartbeats. Heart rate – just like stress – increases over the course of the game, as time runs out and the end of the battle approaches. Their heart rates also spike when their opponent commits an error that will decide the outcome of the game.

But the most significant discovery we made was this: a few seconds before the players made a mistake, their heart rate changed. This means that in a situation with countless options, with a complexity that is similar to that of life itself, the heart panics before making a bad decision. If the players could recognize that, if they were able to listen to what their hearts are telling them, they could perhaps avoid many of their errors.

This is possible because the body and the brain hold the keys to decision-making long before we are consciously aware of those elements; the emotions expressed in our bodies function as an alarm to alert us to possible risks and mistakes. This destroys the idea that intuition belongs to the realm of magic or soothsaying. There is no conflict between hunches and science; in fact, quite the opposite: intuition functions hand in hand with reason and deliberation, fully in the realm of science.

Rational deliberation or hunches?

Once we have discovered that hunches and intuitions are unconscious deliberations we can proceed to a question of more practical relevance. When should we trust our hunches and intuitions and when not? For those questions that matter most to us, should we trust our hunches or our rational deliberations?

The answer is conclusive: it depends. A social psychologist, Ap Dijksterhuis, found, in an experiment that is still generating controversy, that the complexity of a decision is what dictates when it's best to deliberate consciously or act intuitively. Dijksterhuis found that to be the rule both in 'mock' decisions in the lab and in real-life decisions.

> In the laboratory, he constructed a game in which participants had to evaluate two options – for example, two cars – and choose which was preferable in terms of utility. Sometimes, the two alternatives only differed in price. In that case, the decision was simple: the cheaper one was better. Then the problem became progressively more complex, when the two cars varied not only in price, but in petrol consumption, safety, comfort, risk of theft, engine capability and pollution levels.

Dijksterhuis's most surprising discovery was that when there are many elements in play, hunches are more effective than deliberation. The same pattern appears in decisions in the real world. This was observed in an experiment whereby people who had just bought toothpaste – undoubtedly one of life's easier choices – were asked how they had made their decision. A month later, those who had pondered their decisions were more satisfied than those who hadn't. On the other hand, they observed the opposite result when interviewing people who had just bought furniture (a complex decision, with many more variables such as price, size, quality, aesthetic appeal). Just like in the lab, those who thought less chose better.

The methodologies of these experiments are quite different, but the conclusion is the same. When we make a decision by carefully thinking over a small number of elements, we choose better if we take our time. Yet when the problem is complex, in general we make better choices by following our intuition than if we stew over it.

The conscious mind is fairly limited in size and can hold little information. Our unconscious, however, is vast. This explains why when making decisions with few variables in play – price, quality and size of a product, for example – we are best served by thinking it over before acting. In situations where we can mentally evaluate all the elements at the same time, the rational decision is more effective, and therefore better. We also can see why – when there are many more variables in play than our conscious mind can juggle at once – our unconscious, rapid, intuitive decisions are more effective, even when based on approximated calculations.

Sniffing out love

Perhaps the most important and complex decisions that we make are social and emotional. It may seem strange, almost absurd, to decide whom to fall in love with in a deliberate way, by some arithmetic evaluation of arguments for and against that person we feel so drawn to. That's just not how it works. We fall in love for reasons that are generally mysterious and can only be determined sketchily after some time has passed.

At *pheromone parties*, each participant sniffs the clothing that's been worn for a few days by other guests. Based on the odour print that attracts them, they decide whom to approach at the party. Choosing this way seems natural because we associate our sense of smell with intuition, like when we say that 'something smells fishy.' And because we all recognize how evocative the intimate and in-describable scent of our lover's sheets is. But, at the same time, it's weird because, obviously, our sense of smell isn't the most precise of our senses. So it seems fairly likely that someone could be sorely disappointed by the partner their sniffing leads them to, and run off cursing their ridiculous nose.

Claus Wedekind, a Swiss biologist, made a phenomenal experiment out of this game. He had a group of men wear the same T-shirt for several days, with no deodorants or perfumes. Then a series of women smelled the shirts and articulated how pleasurable they found each scent – and, of course, he also did the reverse, having the men sniff the women's well-worn T-shirts. Wedekind wasn't just fishing with this experiment to see what he would find: he had based it on a hypothesis constructed from observing the behaviour of rodents and other species. He was exploring the premise that as far as scent, taste and unconscious preferences were concerned, we are very similar to our inner 'beasts.'

Each individual has a different immune repertoire, which explains, in part, why, when exposed to the same virus, some of us get sick and others don't. We can think of each immune system as a shield. If two shields are placed one on top of the other protecting the same space, they become redundant. However, two shields covering different, contiguous spaces can together protect a larger surface area. The same idea can be transferred – with certain drawbacks that we will ignore for the moment – to the immune repertoire: two individuals with very different immune repertoires give rise to progeny with a more effective immune system.

In rodents, who use their sense of smell much more than we do when choosing a mate, the preference largely follows a simple rule governed by this principle: they tend to choose mates with a different immune repertoire. This was the basis for Wedekind's experiment. He measured each participant's MHC (major histocompatibility complex), a family of genes involved in the differentiation between our own and others' immune systems. And the extraordinary result is that when we judge by our sense of smell, we do so according to the same premise as our rodent cousins: on average, women will be more attracted to the scent of men who have a different MHC. So pheromone parties* promote diversity. At least in terms of immune repertoires.

But this rule has a notable exception. A female mouse's scent preferences invert when she is pregnant. Then she prefers the smell of mice with MHCs that are similar to hers. The simplified, narrative version of this result is that while the search for complementariness can be beneficial when mating, once there is already a baby in the womb it makes sense to remain close to a known nest, among kin, with those who are similar.

Does the same shift in olfactory preference happen with women? It seems plausible since, in the midst of the hormonal revolution that occurs during a woman's pregnancy, her changes in smell and taste perception are among the most distinctive effects. Wedekind studied how olfactory preference changed when a woman is taking birth control pills with steroids that stimulate a very similar hormonal state to pregnancy. Thus it was discovered that, just like in rodents, the result was turned on its head, and the smell of T-shirts worn by men with similar MHCs became more appealing.

This experiment illustrates a much more general concept. Many of our emotional and social decisions are much more stereotypical than we recognize. In general, this mechanism is masked by the mystery of the unconscious and, therefore, we do not perceive the process of deliberation. But it is there, in the underground workings of an

---

* Pheromones are the mediators in a chemical communication system – like our sense of smell – that is specific to species and affects automatic brain functions. Rodents have a specialized system for pheromones called the vomeronasal organ. How (and whether) pheromones work in humans is still disputed and they are usually considered unconscious odours. But this is an imprecise and erroneous definition, since the same molecules of the olfactory system, in small doses, can induce behaviours without conscious awareness. Perhaps the pheromone parties are just odour parties. But, of course, that doesn't sound as glamorous.

apparatus that may have been forged long before we were able even to begin pondering these questions.

In short, decisions that are based on hunches and intuition, which because they are unconscious are often perceived as magical, spontaneous and unfounded, are actually regulated and sometimes markedly stereotypical. According to the mechanical virtues and limitations of our awareness, it seems wise to delegate 'simple' decisions to rational thought and leave the complex ones to our smell, sweat and heart.

Believing, knowing, trusting

When making a decision, in addition to carrying out the chosen option, the brain generates a belief. That is what we perceive as trust or conviction in what we are doing. Sometimes we buy a chocolate bar at a kiosk, certain that it's exactly what we want. Other times we walk away hoping that the chocolate sweetens our frustration at knowing we haven't chosen well. The treat is the same, but the bitter perception of having made a clumsy decision is very different.

We have all, at some point, blindly trusted a decision that later turned out to have been wrong. Or, conversely, in many situations we act without conviction when we actually have all the arguments to be heady with confidence. How is this feeling of trust in our decisions constructed? Why do some people constantly walk around excessively confident, no matter what they do, and others live in doubt?

The scientific study of this trust – or hesitation – turns out to be particularly tempting because it opens up a window on subjectivity; it is not a study of our observable actions but rather of our private beliefs. Which is not to say that it is a minor matter from a purely practical standpoint, since our confidence (or lack thereof) in ourselves and our actions defines our manner of being.

The simplest way to study confidence is asking someone to draw a point on a line where one end represents absolute confidence and the other represents doubt about a decision that's been made. Another way to detect confidence is by asking the decision-makers if they prefer to charge a fixed amount for the decision or make a bet on it to earn more. If they are very confident about the decision they've just made, they'll be inclined to bet (two birds in the bush). If, on the other hand, they are hesitant about their choice, they'll prefer the fixed amount (bird in the hand). These two means of measuring confidence are very consistent; those who display firm conviction on the line model also bet boldly. And the opposite is also true: those who tend to express low confidence in their decisions are not inclined to bet on them.

This parallel between confidence and betting has obvious relevance for daily life. Betting or investing poorly in financial, emotional, professional, political and family questions has a high cost. But this parallel also has scientific consequences. This type of experiment allows us to question our subjectivity in areas that previously seemed to be impossible to broach. When measuring someone's predisposition to betting we are discovering something about confidence perceived by those who cannot express their beliefs in words.

Confidence: flaws and signatures

The way each person constructs confidence is almost like a digital footprint. Some people express confidence with intermediary nuances, and others with extreme doubt or conviction. There are also cultural traits, and the ways certainty is expressed in some parts of Asia differs from how it is expressed in the West.

Almost all of us have witnessed examples where we assign confidence in a fairly imprecise way, such as when we think we did well in an exam and it turns out we failed it. And most of us have also known people who are quite accurate in assessing their own knowledge and therefore have a precise, dependable sense of confidence, knowing when to bet and when not to. Confidence is then a window into one's own knowledge.

The accuracy of one's confidence is a personal trait, similar to height or eye colour. But unlike those physical traits, there is a certain amount of space to change and modify this thought pattern. As an identity trait, it has a signature in the brain's anatomical structure. Those who have more accurate confidence systems have a greater number of connections – measured in density of axons – in a region of the lateral front cortex called Brodmann Area 10, or BA10. Additionally, those with a more precise sense of confidence organize their cerebral activity in such a way that this BA10 region is more efficiently connected to other cortical structures in the brain, such as the angular gyrus and the lateral frontal cortex.

This difference in functional brain connectivity between those who have an accurate sense of confidence and those who don't is only observed when a person turns their attention inwards – for example, by focusing on their breathing – and not when their attention is focused outwards on the external world. This establishes a bridge between two variables that were seemingly scarcely related: our sense of confidence and our knowledge of our own body. What they have in common is that they both lead our thoughts inwards. And that suggests that a natural way of improving accurate confidence in our decision-making system is learning to observe and focus on our own body states.

To sense whether we trust ourselves or not, our brain uses endogenous variables. For instance, it will sense lack of confidence if we sweat, stutter, lower our gaze or express other bodily signs of doubt. These body signs, which we use to sense confidence in others, also allow us to sense that about ourselves.

The nature of optimists

When the balance between doubt and certainty applies to outcomes in the unknowable future, our sense of confidence divides us into optimists or pessimists. Optimists are sure they will make every shot, win every big game, never lose their job and can have unprotected sex or drive recklessly because, after all, they are immune to the risks. What's mysterious is how optimism survives despite knock-backs and the evidence to the contrary we receive each and every day. The solution to this conundrum in an optimist's brain is selective forgetting. Every Monday, like every 1 January, is filled with repeated promises; every love is the love of our lives, and this year we are absolutely going to win the championship. Each of these affirmations completely ignores the fact that there have been plenty of other Mondays and plenty of other disappointments. Are we really so blind to the evidence? What are the mechanisms in our brains that bring about this fundamentalist adherence to optimism? And what do we do with this persistent optimism while accepting that it's based on an illusion?

One of the most common models of human learning – now delegated chiefly to robotic and artificial intelligence – is the prediction error. It is simple and intuitive. The first premise is that each action we realize, from the most mundane to the most complex, is built on an internal model, a sort of simulated prelude to what will happen. For example, when we greet someone in a lift we are presuming that there will be a positive response from that person. If the response is different from what we are expecting – exaggeratedly warm or coldly reluctant – we are surprised.

Prediction error expresses the difference between what we expect and what we actually observe, and that is codified in a neuronal circuit in the basal ganglia, which generates dopamine. Dopamine is a neurotransmitter whose functions include being a messenger of surprise when travelling into various brain structures. The dopaminergic signal recognizes the dissonance between what is predicted and what is found, and it is the fuel of learning because circuits irrigated with dopamine become malleable and predisposed to change. In the absence of dopamine, neuronal circuits are generally rigid and not very malleable.

The cyclical renewal of our hopes, every Monday and every New Year's Eve, forces us to hack into this learning system. If the brain didn't generate a signal of dissonance when reality is worse than what we were expecting, we would indefinitely renew our hopes. Is that what happens? And if so, how? Is this the optimist's secret gift?

All these questions are answered in unison by a relatively simple experiment conducted by a British neuroscientist, Tali Sharot. In it, she asks people to estimate the probability that various unfortunate events would occur. What is the likelihood of dying younger than sixty years old? What about developing a degenerative disease? What about having a car accident?

A large majority of those asked assume that the odds of something bad happening to them are lower than what statistics suggest. Which is to say, when trying to evaluate our risks – flying in a plane and urban violence are clear exceptions – we are almost all decidedly optimistic.

But the most interesting thing is what happens when our beliefs do not coincide with reality. For example, in the experiment, participants were asked to estimate the likelihood of suffering from cancer and later on were told that the average likelihood of someone like them doing so is close to 30 per cent.

According to the model of prediction error, people should use this information to modify their beliefs. And that is exactly what happens when most people discover that things are better than they supposed. Participants who believed that their likelihood of suffering from cancer was worse than in reality adjusted their outcomes to a value very close to the real one. For instance, those who believed that their likelihood was 50 per cent responded in subsequent interviews with values around 35 per cent, which is quite close to the real value of 30 per cent.

But – and herein lies the key – those who believed that their probability of suffering cancer was less than in reality (for instance those responding that it was 10 per cent) changed their beliefs very little. When asked again in subsequent interviews and after having heard the bad news that the likelihood of suffering from cancer was in fact 30 per cent, they adjusted their estimates by a mere 1 or 2 per cent (to 11 or 12 per cent). That is to say, the adjustment those people make is much less – almost nil – when they discover that the truth is worse than they imagined it to be.

And what happens meanwhile in the brain? Every time we discover a desirable or beneficial piece of information, a group of neurons in a small part of the left prefrontal cortex called the inferior frontal gyrus is activated. On the other hand, when we receive undesired information, another group of neurons activates in the homologous region on the right hemisphere. A sort of equilibrium between the good and bad news is established between these brain regions. But this equilibrium has two traps: the first is that it focuses much more on good news than on bad, which, on average, creates a tendency towards optimism; and secondly – and most interestingly – the bias in the balance varies from person to person, thus revealing the machinery behind optimism.

The activation of neurons in the frontal gyrus of the left hemisphere is similar in everyone when we discover that the world is better than we had thought. On the other hand, the activation of the frontal gyrus of the right hemisphere varies widely from one individual to the next when we find out that the world is worse than we believed it to be. In more optimistic people, this activation is minimized, as if they literally turned a blind eye on bad news. In more pessimistic people, the opposite happens; the activation is amplified, accentuating and multiplying the impact of that negative information. Here is the biological recipe that separates the optimists from the pessimists: it is not their capacity to value what's good but rather their ability to ignore and forget what's bad.

Many mothers, for example, have only a vague recollection of the pain they experienced during childbirth. That selective forgetting eloquently illustrates the mechanism of optimism. If the pain were much more present in their memory, perhaps we would see many more only children. Something similar happens among newlyweds; none of them believe they will ever divorce. Yet between 30 and 50 per cent of them will, according to statistics that vary based on time and place. Of course, the moment when they swear eternal love – whatever is meant by love and eternity – isn't the most appropriate time for statistical reflections on human relationships.

The costs and benefits of excessive or insufficient optimism are pretty tangible. There are intuitive reasons to encourage a certain amount of naïve optimism, since it turns out to be a driving force behind action, adventure and innovation. Without optimism we would never have landed on the moon. Optimism is also associated in a fairly generic way with better health and a more satisfying life. So we could think of optimism as a sort of little insanity that pushes us to do things that we otherwise wouldn't do. Its flip side, pessimism, will lead to inaction and, in its chronic version, to depression.

But there are also good reasons to temper excessive optimism when it encourages risky and unnecessary decisions. Conclusive statistics associating the risk of car accidents with being inebriated, using a mobile phone or not wearing a seatbelt continue to pile up. Optimists know these risks but act as if they are immune to them. They feel they are exempt from the statistics and this, of course, is false: if we were all exceptions, the rule would not exist. This expansive optimism – which usually does not consider itself as such – can lead to fatal yet avoidable consequences.

Odysseus and the consortium we belong to

A much more mundane example of excessive optimism is our waking up each day. Often bedtime is filled with promises about the next morning: we plan to get up much earlier than usual to, for example, exercise. That intention is built on a genuine desire and on an expectation of some value to us, such as increased health and fitness. But, except for larks, the panorama the next day is a very different one. The person who made the decision the night before to get up early disappears by next morning. At 7 a.m. we are somebody different altogether, overcome by sleepiness and the strictly hedonistic pleasure of staying in bed.

The outlines of identity are blurred. Or, to put it more precisely, each of us makes up a consortium of identities that are expressed in different, sometimes contradictory, ways in varying circumstances. The disassociation between various members of the consortium has two clear projections: one that is hedonistic and bold, that ignores the risks and future consequences (the optimist), and another that ponders those risks and consequences (the pessimist). This dynamic is particularly exacerbated in two quite different scenarios: in certain neurological and psychiatric pathologies, and in adolescence.

The predisposition to ignore risk grows with the activation of the nucleus accumbens in the limbic system, which corresponds to the perception of hedonistic pleasure. In fact, in an experiment that shocked some of his colleagues at the Massachusetts Institute of Technology, Dan Ariely recorded this in a detailed, quantitative way with regard to a precise aspect of pleasure: sexual arousal. He found that the more excited people get, the more predisposed they are to doing things that they would otherwise consider aberrant or unacceptable. Such things of course included taking the risk of having unprotected sex with strangers.

Adolescence is a period plagued with excessive optimism and exposure to risky situations. This happens because the brain's development, like the body's, is not homogeneous. Some cerebral structures develop very quickly and mature within the first few years of life, while others are still immature when we become teenagers. One popular neuroscientific myth is that adolescence is a time of particular risk because of the immaturity of the prefrontal cortex, a structure that evaluates future consequences and coordinates and inhibits impulses. However, the later development of the control structure in the frontal cortex cannot explain per se the spike in risk predisposition recorded during the teenage years. In fact, children, whose prefrontal cortex is even more immature, expose themselves to less risk. What is characteristic of adolescence is the relative immaturity of prefrontal cortex development – and as a result, the ability to inhibit or control certain impulses – with a consolidated development of the nucleus accumbens.

The naïve clumsiness of those teenage years, in a body that is growing more than its capacity to control itself, can be seen as a reflection of the adolescent cerebral structure. Understanding this, and taking into account the uniqueness of this time in our lives, can help us to empathize and, as a result, engage in a dialogue more effectively with teenagers.

This understanding of the brain structure is also relevant for making public decisions. For example, in many countries there is debate surrounding whether teenagers should be allowed to vote. These debates would benefit from taking into account an informed view of the development of reasoning and the process of decision-making during adolescence.

The work done by Valerie Reyna and Frank Farley on risk and rationality in teenagers' decision-making shows that, even when they don't have good control of their impulses, teenagers are intellectually indistinguishable from adults in terms of rational thought. Which is to say, they are capable of making informed decisions about their future despite the fact that they struggle, more than an adult would, to rein in their impulses in emotionally charged states.

But, of course, we don't need a biologist to tell us that we alternate between reason and impulse, and that our impulsivity shows up in the heat of the moment even beyond our teenage years. This is expressed in the myth of Odysseus and the Sirens, which also gives us perhaps the most effective solution for dealing with this consortium that comprises our identity. When heading off on his voyage home to Ithaca, Odysseus asks his sailors to tie him to the boat's mast so that he won't act on the inevitable temptation to follow the Sirens' song. Odysseus knows that in the heat of the moment, the craving will be irresistible,* but instead of cancelling his voyage he decides to make a pact with himself, binding together his rational self with his anticipated future impulsive one.

The analogies with our daily life are often much more banal; for

---

* 'I can resist anything except temptation', as Oscar Wilde said.

many of us, our mobile phones ring out with the contemporary version of the Sirens' song, virtually impossible to ignore. To such an extent that, although we know the clear risks of answering a text while at the wheel, we do it even when the message is something completely irrelevant. Ignoring the temptation to use our phone while driving seems difficult, but if we leave it somewhere inaccessible – such as in the boot of the car – we, like Odysseus, can force our rational thinking to control our future recklessness.

Flaws in confidence

Our brain has evolved mechanisms to ignore – literally – certain negative aspects of the future. And this recipe for creating optimists is just one of the many ways the brain produces a disproportionate sense of confidence. Studying human decisions in the social and economic problems of daily life, Daniel Kahneman, a psychologist and Nobel Prize laureate in Economics, identified two archetypal flaws in our sense of confidence.

The first is that we tend to confirm that which we already believe. That is to say, we are generally headstrong and stubborn. Once we believe something, we look to nourish that prejudice with reaffirming evidence.

One of the most famous examples of this principle was discovered by the great psychologist Edward Thorndike when he asked a group of military leaders what they thought about various soldiers. The opinions dealt with different aptitudes that included physical traits, leadership abilities, intelligence and personality. Thorndike proved that the evaluation of a person mixes together abilities that, on the face of them, have no relationship to each other. That was why the generals rated the strong soldiers as intelligent and good leaders, although there is no necessary correlation between strength and intelligence.*

---

* That was not only how the generals thought, but how we all do in general.

Which is to say that when we evaluate one aspect of a person, we do so under the influence of our perception of their other traits. And this is called the *halo effect*.

This flaw of the decision-making mechanism is not only pertinent to daily life but also in education, politics and the justice system. No one is immune to the halo effect. For example, when faced with an identical group of conditions, judges are more lenient with people who are more attractive. This is an excellent example of the halo effect and the deformations it causes: those who are lovely to look at are viewed as good people. Of course, this same effect weighs on the *free and fair* mechanism of democratic elections. Alexander Todorov showed that a brief glance at the face of two candidates allows one to predict the winner with striking precision – close to 70 per cent, even without data on the candidates' history, thoughts and deeds, or their electoral platforms and promises.

The confirmation bias – the generic principle from which the halo effect derives – cuts reality down so we see only what is coherent with what we already believe to be true. 'If she looks competent, she'll be a good senator.' This inference, which ignores facts pertinent to the assessment and is based entirely on a first impression, turns out to be much more frequent that we realize or will admit to in our day-to-day decisions and beliefs.

In addition to the confirmation bias, a second principle that inflates confidence is the ability to completely ignore the variance of data. Think about the following problem: a bag holds 10,000 balls, you take out the first one and it's red, you take out a second one and it's red too. You take out the third and fourth, and they are red as well. What colour will the fifth one be? Red, of course. Our confidence in that conclusion far outweighs the statistical probability. There are still 9,996 balls in the bag. As Woody Allen says, 'Confidence is what you have before you understand the problem.' To a certain extent, confidence is ignorance.

Postulating a rule based on just a few cases is both a virtue and a vice of human thought. It is a virtue because it allows us to identify rules and regularities with consummate ease. But it is a vice because it pushes us towards definitive conclusions when we have barely observed a tiny slice of the reality. Kahneman proposed the following mental experiment. A survey of 200 people indicates that 60 per cent would vote for a candidate named George. Very shortly after finding out about that survey, the only thing we remember is that 60 per cent would vote for George. The effect is so strong that many people will read that and think that I wrote the same thing twice. The difference is the size of the sample. In the first phrasing, the case explicitly states that it is the opinion of only 200 people. In the second, that information has disappeared. This is the second filter that distorts confidence. In fact, in formal terms, a survey showing that out of 30 million people 50.03 per cent would vote for George would be much more decisive, but the belief system in our brains mostly makes us forget to weigh in whether the data comes from a massive sample or whether we are dealing with three balls in a bag of 10,000. As the recent 'Brexit' outcome in the UK or the Donald Trump vs Hillary Clinton election show, often, in the build-up to an election, the pollsters forget this basic rule of statistics and draw firm conclusions based on a strikingly small and often biased amount of data.

In short, the confirmatory effect and variance blindness are two ubiquitous mechanisms that, in our minds, allow us to base opinions on just a small portion of the coherent world while ignoring an entire sea of noise. The direct consequence of these mechanisms is inflated confidence.

A vital question in understanding and improving our decision-making is to explore if these confidence flaws are native to complex social decisions or if they are seen throughout the vast spectrum of decision-making. Ariel Zylberberg, Pablo Barttfeld and I set out to solve this mystery by studying extremely simple decisions, such as which is the brighter of two points of light. We found that the principles which inflate confidence in social decisions, such as the confirmatory effect or ignoring variance, are traits that persist even in the simplest perceptual decisions.

It is a common trait of our brains to generate beliefs that are more optimistic than actual data suggests. This was confirmed by a series of studies recording the neuronal activity in different parts of the cerebral cortex. It was consistently observed that our brains – and the brains of many other species – are constantly mixing sensory information from the outside world with our own internal hypotheses and conjectures. Even our vision, the brain function we imagine to be most anchored to reality, is filled with illusions. Vision doesn't function passively depicting reality like a camera, but rather more like an organ interpreting and constructing detailed images based on limited and imprecise information. Even in the first processing station in the visual cortex, neurons respond according to a conjunction of information received by the retina and information received by other parts of the brain – parts that codify memory, language, sound – which establish hypotheses and conjectures about what is being seen.

Our perception always involves some imagination. It is more similar to painting than to photography. And, according to the confirmation effect, we blindly trust the reality we construct. This is best witnessed in visual illusions, which we perceive with full confidence, as if there were no doubt that we are portraying reality faithfully. One interesting way of discovering this – in a simple game that can be played at any moment – is the following. Whenever you are with another person, ask him or her to close their eyes, and start asking questions about what is nearby – not very particular details but the most striking elements of the scene. What is the colour of the wall? Is there a table in the room? Does that man have a beard? You will see that the majority of us are quite ignorant about what lies around us. This is not so puzzling. The most extraordinary fact is that we completely disregard this ignorance.

Others' gazes

Both in everyday life and in formal law we judge others' actions not so much by their consequences but by their determining factors and motivations. Even though the consequence may be the same, it is morally very different to injure a rival on a playing field through an unfortunate, involuntary action than through premeditation. Therefore, in order to be able to decide whether the player acted with bad intentions, just observing the consequences of their actions is not enough. We must put ourselves in their place and see what happened from the victim's perspective. Which is to say, we have to employ what is known as the *theory of mind*.

Let us consider two fictional situations. Joe picks up a sugar bowl and serves a spoonful into his friend's tea. Before he did so, someone switched the sugar for a poison of the same colour and consistency. Of course, Joe didn't know that. His friend drinks the tea and dies. The consequences of Joe's action are tragic. But was what he did wrong? Is he guilty? Almost all of us would say no. In order to arrive at that conclusion, we put ourselves in his shoes, recognizing what he knows and doesn't know, and seeing that he had no intention of hurting his friend. Not only that, but in most people's minds he was not even negligent in any way. Joe is a good guy.

Same sugar bowl, same place. Peter takes the bowl and replaces the sugar with poison because he wants to kill his friend. He spoons the poison into his friend's tea but it has no effect on him, and his friend walks away alive and kicking. In this case, the consequences of Peter's action are innocuous. However, we almost all believe that Peter did the wrong thing, that his action is reprehensible. Peter is a bad guy.

The theory of mind is the result of the articulation of a complex cerebral network, with a particularly important node in the right temporoparietal junction. As its name suggests, it is found in the right hemisphere, between the temporal and parietal cortices, but its location is the least interesting thing about it. Cerebral geography is less important than the fact that a function's location in the brain can be a window to inferring the causal relationships in its workings.*

---

* Coils that generate magnetic fields can be used to silence or stimulate a

If our right temporoparietal junction were to be temporarily silenced, we would no longer consider Joe and Peter's intentions when judging their actions. If that region of our brains is not functioning as it should, we would believe that Joe did wrong (because he killed his friend) and that Peter did right (because his friend is in perfect health). We wouldn't take into consideration that Joe didn't know what was in the sugar bowl and that Peter had failed to carry out his macabre plan only through chance. These considerations require a precise function, the theory of mind, and without it we lose the mental ability to separate the consequences of an action from its network of intentions, knowledge and motivations.

This example, demonstrated by Rebecca Saxe, is proof of a concept that goes beyond the theory of mind, morality and judgement. It indicates that our decision-making machinery is composed of a combination of pieces that establish particular functions. And when the biological support of those functions is disarmed, the way we believe, form opinions and judge changes radically.

More generally, it suggests that our notion of justice does not result from pure and formal reasoning, but that instead it is conceived in a particular state of the brain.*

But there is no need in fact to have very sophisticated brain-stimulating devices to prove this concept. The common image of the reality of justice being 'what the judge ate for breakfast' seems in fact to be quite true. The percentage of favourable rulings by US judges drops dramatically during the morning, then peaks abruptly after the lunch break to drop again substantially in the next session. This study of course cannot factor out the many variables that change between breaks such as glucose, or fatigue, or accumulated stress.

---

region of the cerebral cortex at a particular moment in time. For example, this technique could be used to stimulate the Broca area, which coordinates language articulation, and induce an unstoppable verbosity.

* As Groucho Marx famously said, 'Those are my principles, and if you don't like them . . . well, I have others.'

But it shows that simple extraneous factors which condition the state of the judge's brain have a strong influence on the outcome of court decisions.

The inner battles that make us who we are

Moral dilemmas are hypothetical situations taken to an extreme that help us reflect on the underpinnings of our morality. The most famous of them is the 'trolley problem', which goes like this:

> You are on a tram without brakes that is travelling along a track where there are five people. You are well acquainted with its functioning and know without a shadow of a doubt that there is no way to stop it, and that it will run over those five people. There is only one option. You can turn the wheel and take another track where only one person will be run over.

Would you turn the wheel? In Brazil, Thailand, Norway, Canada and Argentina, almost everyone – young or old, liberal or conservative – chooses to turn it on the basis of a reasonable, utilitarian calculation. The choice seems simple: five deaths or one? Most people across the world choose to kill one person and save five. Yet experiments show that there is a minority of people who consistently decide not to turn the wheel.

The dilemma consists in doing something that will provoke the death of one person or not doing anything to keep five people from dying. Some people could reason that fate had already chosen a path and that they shouldn't play God and decide who dies and who lives, even when the maths favours that choice. They reason that we have no right to intervene, bringing about the death of somebody who would have been fine if not for our action. We all have a different judgement of the responsibility for action or inaction. It is a universal moral intuition that is expressed in almost every legal system.

Now, another version of the dilemma:

You are on a bridge from which you see a tram hurtling down a track where there are five people. You are completely sure that there is no way to stop it and that it will run over those five people. There is only one alternative. On the bridge there is a large man sitting on the railing watching the scene. You know for certain that if you push him, he will die but he will also make the trolley go off the tracks and save the other five people.

Would you push him? In this case, almost everyone chooses not to. And the difference is perceived in a clear, visceral way, as if it were decided by our bodies. We don't have the right deliberately to push someone to save someone else's life. This is supported by our penal and social system – both the formal one and the judgements of our peers: neither would consider these two cases to be equal. But let's forget about that factor. Let's imagine that we are alone, that the only possible judgement is our own conscience. Who would push the man from the bridge and who would turn the wheel? The results are conclusive and universal: even completely alone, with no one watching, almost all of us would turn the wheel and almost no one would push the man from the bridge.

In some sense, both dilemmas are equivalent. Seeing this is not easy because it requires going against our intuitive body signals. But from a purely utilitarian perspective, from the motivations and consequences of our actions, the dilemmas are identical. We choose to act in order to save five at the expense of one. Or we choose to let fate take its course because we feel that we do not have the moral right to intervene, condemning someone who was not destined to die.

Yet from another perspective, both dilemmas are very different. In order to exaggerate the contrasts between them, we present an even more far-fetched dilemma.

You are a doctor on an almost deserted island. There are five patients, each one with an illness in a different organ that can be solved with a single transplant that you know will undoubtedly leave them in perfect health. Without the transplant they will all die. Someone shows up in the hospital with the flu. You know you can anaesthetize them and use their organs to save the other five. No one would know and it is only a matter of your own conscience being the judge.

In this case, the vast majority of people presented with the dilemma would not take out the organs to save the other five, even to the point of considering the possibility aberrant. Only a few very extreme pragmatists choose to kill the patient with the flu. This third case again shares motivations and consequences with the previous dilemmas. The pragmatic doctor works according to a reasonable principle, that of saving five patients when the only option in the universe is one dying or five dying.

What changes in the three dilemmas, making them progressively more unacceptable, is the action one has to take. The first act is turning a wheel; the second, pushing someone; and the third, cutting into people with a surgical knife. Turning the wheel isn't a direct action on the victim's body. Furthermore, it seems innocuous and involves a frequent action unconnected to violence. On the other hand, the causal relationship between pushing the man and his death is clearly felt in our eyes and our stomachs. In the case of the wheel, this relationship was only clear to our rational thought. The third takes this principle even further. *Slaughtering* a person seems and feels completely impermissible.

The first argument (five or one) is utilitarian and rational, and is dictated by a moral premise, maximizing the common good or minimizing the common evil. This part is identical in all three dilemmas. The second argument is visceral and emotional, and is dictated by an absolute consideration: there are certain things that are just not done. They are morally unacceptable. This specific action that needs to be done to save five lives at the expense of one is what distinguishes the three dilemmas. And in each dilemma we can almost feel the setting in motion of a decision-making race between emotional and rational arguments, *à la* Turing, in our brain. This battle that invariably occurs in the depths of each of us is replicated throughout the history of culture, philosophy, law and morality.

One of the canonical moral positions is deontological – this word derives from the Greek *deon*, referring to obligation and duty – according to which the morality of actions is defined by their nature and not their consequences. In other words, some actions are intrinsically bad, regardless of the results they produce.

Another moral position is utilitarianism: one must act in a way that maximizes collective wellbeing. The person who turns the wheel, or pushes the man, or slices open the flu sufferer would be acting according to a utilitarian principle. And the person who does not do any of those actions would be acting according to a deontological principle.

Very few people identify with one of those two positions to the extreme. Each person has a different equilibrium point between the two. If the action necessary to save the majority is very horrific, deontology wins out. If the common good becomes more exaggerated – for example, if there are a million people being saved instead of five – utility moves into the foreground. If we see the face, expression or name of the person to be sacrificed for the majority – particularly when it is a child, a relative or someone attractive – deontology again has more weight.

The race between the utilitarian and the emotional is waged in two different nodes of the brain. Emotional arguments are codified in the medial part of the frontal cortex, and evidence in favour of utilitarian considerations is codified in the lateral part of the frontal cortex.

Just as one can alter the part of the brain that allows us to understand another person's perspective and hack into our ability to use the theory of mind, we can also intervene in those two cerebral systems in order to inhibit our more emotional part and foster our utilitarianism. Great leaders, such as Churchill, usually develop resources and strategies to silence their emotional part and think in the abstract. It turns out that emotional empathy can also lead us to commit all sorts of injustices. From a utilitarian and egalitarian perspective of justice, education and political management, it would be necessary to detach oneself – as Churchill did – from certain emotional considerations. Empathy, a fundamental virtue in concern for our fellow citizens, fails when the goal is to act in the common good without privilege and distinctions.

In everyday life there are very simple ways to give more weight to one system or the other. One of the most spectacular was proven by my Catalan friend Albert Costa. His thesis is that when making the cognitive effort to speak a second language we place ourselves in a mode of cerebral functioning that favours control mechanisms. As such, it also favours the medial part of the frontal cortex that governs the utilitarian and rational system of the brain. According to this premise, we could all change our ethical and moral stance depending on the language we are speaking. And this does, in effect, occur.

Albert Costa showed that native Spanish-speakers are more utilitarian when speaking English. If a Spanish-speaker were to be posed with the man-on-the-bridge dilemma in a foreign language, in many cases he or she would be more willing to push him. The same thing happens with English-speakers: they become more pragmatic when evaluating similar dilemmas in Spanish. For a native English-speaker it is easier to push a man in Spanish.

Albert proposed a humorous conclusion to his study, but one which surely has some truth to it. The battle between the utilitarian and the emotional is not exclusive to abstract dilemmas. Actually, these two systems are invariably expressed in almost all of our deliberations. And, many times, in the safety of our homes more ardently than anywhere else. We are in general more aggressive, sometimes violently and mercilessly, with those we love most. This is a strange paradox in love. Within the trust of an unvarnished, unprejudiced relationship with vast expectations, jealousy, fatigue and pain, sometimes irrational rage emerges. The same argument between a couple that seems unbearable when we are living through it becomes insignificant and often ridiculous when seen from a third-person perspective.

Why are they fighting over something so stupid? Why doesn't one or the other simply give in so they can reach an agreement? The answer is that the consideration is not utilitarian but rather capriciously deontological. The deontology threshold drops precipitously and we are not willing to make the slightest effort to resolve something that would alleviate all the tension. Clearly we would be better off being more rational. The question is, how? And Albert, half joking and half serious, suggests that the next time we are fighting with our significant other, we should switch to Spanish (or any other non-native language). This would allow us to bring the argument into a more rational arena, one less burdened by visceral epithets.

The moral balance is complicated. In many cases, acting pragmatically or in a utilitarian manner requires detaching ourselves from strong emotional convictions. And it implies (most often implicitly) assigning a value (or a prize) to issues that, from a deontological perspective, it seems impossible to rationalize and convert into numbers.

Let's perform a concrete mental experiment to illustrate this point. Imagine that you are going to be late for an important meeting. You are driving and right after you have crossed a railway line you realize that the warning signs at the level crossing are not working. You feel lucky that no train was passing when you drove across. But you understand that with the traffic due to get heavier someone will be hit by a train and most likely die. You then call 999 to inform the emergency services, but at the same time you realize that, if you don't make the call, the fatal accident will close the streets just behind you and prevent traffic coming from various places in the city. And with that, you will make it on time to work. Would you hang up and let someone die to gain a few minutes and make it on time to your meeting? Of course not. The question seems absurd.

Now imagine that it is five of you in the same car coming together. You are the only one who realizes that the warning signals are not working – maybe because as a child you were fascinated by level crossings. Same question and surely same answer. Even if no one would know 'your sin', you would make the call and prevent the accident. It does not matter if it is one, five, ten or one million people arriving late, the answer is the same. More and more people being late wouldn't add up to the value of one life. And this principle seems to be quite general. Most of us have a strong conviction that, regardless of the dilemma, an argument about life and death would trump all other considerations.

However, we may not live up to this conviction. As absurd as the previous dilemma seems, similar considerations are made daily by each driver and by policy-makers who regulate traffic in major cities. In Great Britain alone about 1,700 people die as a result road accidents. And even if this is a dramatic decrease from the number in the 1980s (close to 6,000), these numbers would be way lower if traffic speed were further restricted to, say, 25 mph. But this, of course, has a cost. It would take us twice the time to get to work.*

If we forget the cases in which fast driving actually saves lives, as with ambulances, it becomes clear that we are all making an unconscious and approximate comparison that has time, urgency, production and work on one side of the equation, and life and death on the other.

Establishing rules and principles for morality is a huge subject that is at the heart of our social pact and, obviously, goes far beyond any analysis of how the brain constructs these judgements. Knowing

---

* From *The Simpsons*:
Homer: My name's Homer Simpson, I'd like to sign up for something.
Mrs Blumenstein: Well, we have an opening on the debate team.
Homer: Debate, like, arguing?
Mrs Blumenstein: Yes.
Homer: I'll take THAT, you DING POT! Just warming up, Mrs Blumenstein.
Mrs Blumenstein: This year's topic is 'Resolved: the national speed limit should be lowered to fifty-five miles per hour'.
Homer: Fifty-five? That's ridiculous! Sure, it'll save a few lives, but millions will be late!

that certain considerations make us more utilitarian can be useful for those who are struggling to behave more in that way, but it has no value in justifying one moral position over another. These dilemmas are only helpful in getting to know ourselves better. They are mirrors that reflect our reasons and our demons so that, eventually, we can have them more at our disposal and not simply silently dictating our actions.

The chemistry and culture of confidence

Ana is seated on a bench in a square. She is going to play a game with another person, chosen at random among the many people in the square. They don't know each other, and they do not see each other or exchange a single word. It is a game between strangers.

The organizer explains the rules of the game, and gives Ana fifty dollars. It is clearly a gift. Ana has only one option: she can choose how to divide the money with the other person, who will remain unaware of her decision. What will she do?

The choices vary widely, from the altruists who fairly divide the money to the egotists who keep it all. This seemingly mundane game, known as the 'Dictator Game', became one of the pillars of the intersection between economics and psychology. The point is that most people do not choose to maximize their earnings and share some of the tokens even when the game is played in a dark room where there is no record of the decision made by the *dictator*. How much is offered depends on many variables that define the contours of our predisposition to share.

Just to name a few: women are more generous and share more tokens independently of their monetary values. Instead, men tend to be less generous and even less as the value of the tokens increases. Also, people behave more generously when under the gaze of real human eyes. What is even more surprising is that just displaying artificial eye-like images on the screen on which players are making their choice makes them share more tokens. This shows that even minimal and implicit social cues can shape our social behaviour. Names also matter. Even when playing with recipients that they have never met, and that they do not see, dictators share more of their tokens when the recipient's name is mentioned. And, on the contrary, a more selfish behaviour can be primed if the game is presented using a market frame of buyers and sellers. Last, ethnicity and physical attractiveness also dictate the way people share, but in a more sophisticated manner. In a seminal study conducted in Israel, Uri Gneezy showed that in a trust game which involved reciprocal negotiations, participants transferred more money when the recipient was of Ashkenazic origin than when they were of Eastern origin. This was true even when the donor was of Eastern origin, showing that they discriminate against their own group. However, in the dictator game, players shared similarly with recipients of both origins. Gneezy's interpretation is that discrimination is the result of ethnic stereotypes (the belief that some groups are less trustable) and not a reflection of an intrinsic taste for discrimination (the desire to harm or offer less to some groups per se). Attractiveness also modulates sharing behaviour in a more complicated way. Attractive recipients tend to receive more, but this seems to depend a lot on specific conditions of how the game is played. One study found that differences based on attractiveness are more marked when the dictators can see the recipient but also listen to them. And the list is much longer. The point is that there are a great number of variables, from very sophisticated social and cultural constructs to very elementary artificial features that shape, in a very predictable manner, our sharing behaviour. And, most often, without us knowing anything about this.

Eva takes part in another game. It also begins with a gift, of fifty dollars that can be shared at will with a stranger named Laura. In this game, the organizers will triple what Eva gives Laura. For example, if she decides to give Laura twenty dollars, Eva will be left with thirty and Laura will get sixty. If she decides to give Laura all of it, Eva will have nothing and Laura will get 150. At the same time, when Laura gets her money, she has to decide how she wants to share it with Eva. If the two players can come to an agreement, the best strategy would be for Eva to give her all the money and then Laura would split it equally. That way they would each get seventy-five dollars. The problem is that they don't know each other and they aren't given the opportunity to make that negotiation. It is an act of faith. If Eva decides to trust that Laura will reciprocate her gracious action, she should give her all the money. If she thinks that Laura will be stingy, then she shouldn't give her any. If – like most of us – she thinks a little of both, perhaps she should act Solomonically and keep a little – a sure thing – and invest the rest, accepting the social risk.

This game, called the 'Trust Game', directly evokes something we have already covered in the realm of optimism: the benefits and risks of trust. Basically, there are plenty of situations in life in which we would do better by trusting more and cooperating with others. Seen the other way around, distrust is costly, and not only in economic decisions but also in social ones – surely the most emblematic example being in couple relationships.

The advantage of taking this concept to its most minimal version in a game/experiment is that it allows us to exhaustively investigate what makes us trust someone else. We had already guessed some elements. For example, many players in the experiment often find a reasonable balance between trusting and not exposing themselves completely. In fact, the first player usually offers an amount close to half. And trusting the other person depends on the similarities between the players, in terms of accent, facial and racial features, etc. So again we see the nefarious effects of a morality based on superficiality. And what a player offers also depends on how much money is at stake. Someone who may be willing to offer half when playing with ten quid might not do the same when playing with 10,000. Trust has a price.

In another variant of these games, known as the 'Ultimatum Game', the first player, as always, must decide how to distribute what they have been given. The second player can accept or reject the proposal. If they reject it, neither of them gets anything. This means that the first player has to find a fair balance point that is usually a little above nothing and a little below half. Otherwise, both players lose.

Bringing this game to fifteen small communities in remote parts of the planet, and in search of what he called *homo economicus*, the anthropologist Joseph Henrich discovered that cultural forces establish quite precise rules for this type of decision. For example, in the Au and Gnau peoples of Papua New Guinea, many participants offered more than half of what they received, a generosity rarely observed in other cultures. And, to top it all off, the second player usually rejected the offer. This seems inexplicable until one understands the cultural idiosyncrasy of these Melanesian tribes. According to implicit rules, accepting gifts, even unsolicited ones, implies a strict obligation to repay them at some point in the future. Let's just say that accepting a gift is like taking on some sort of a mortgage.*

Two large-scale studies, one carried out in Sweden and the other in the United States, using twins both monozygotic (identical) and dizygotic (fraternal ones, whose genomes are as different as any other siblings), show that individual differences in generosity seen in the trust game also have a genetic predisposition. If a twin tends to be very generous, in most cases their identical twin will also be. And the opposite is also true: if one decides to keep all the money, there is a high likelihood that the identical twin will do the same. This relationship is

---

* This also happens in our society, in the case of someone who prefers to pay for something rather than take on the commitment and debt involved in accepting a gift. The most exaggerated example of this is expressed by the Italian philosopher Roberto Esposito as the idea that life is a gift that commits us for ever.

found to a lesser extent in dizygotic twins, which allows us to rule out that this similarity is merely a result of having grown up together, side by side, in the same home. That, of course, doesn't contradict what we already saw and intuited: that social and cultural differences influence cooperative behaviour. It merely shows that they aren't the only forces that govern generosity.

Finding a genetic footprint in the predisposition to trust and cooperate leads to a somewhat uncomfortable question. What chemical, hormonal and neuronal states make a person more predisposed to trust others? As with olfactory preferences, a natural starting point for studying the chemistry of cooperation is examining what happens in other animals. And a likely chemical candidate emerges: oxytocin, a hormone that modulates brain activity and plays a key role in the predisposition to social bonding. A player who inhales oxytocin plays the trust game much more generously than a player who inhales a placebo.

Oxytocin is involved in parenting behaviour. In fact, it plays a primary role in the process of activating the uterus during childbirth, which explains its etymology: from the Greek, *oxys*, which means 'quick', and *tokos*, 'birth'. It is also released by sucking on the nipple, which facilitates nursing. But oxytocin not only predisposes the body for motherhood: it also prepares mothers' characters for the huge feat they are about to undertake. Virgin sheep, when receiving oxytocin, behave maternally with the lambs of others as if they were their own. They become great mothers. And vice versa, mother sheep, when given antagonistic substances that block the action of oxytocin, lose their typical maternal behaviours and neglect their offspring. So oxytocin was established as the molecule of maternal love and, more generically, of all love.

Since then a large body of research has shown in one experiment after another that administering a single dose of oxytocin improved different aspects of social cognition: trust, emotional recognition, the ability to direct and sustain gaze at others, understanding, cooperating, and reasoning about high-level social interactions. Oxytocin emerged in the media, with some reason, as the holy grail for empathy, social interactions, emotions and love. Could we simply irrigate oxytocin and make our world a better one? Would dreams of peace, trust, bonding, and a more just and caring society be solved by increasing the dose of a hormone?

The oxytocin hype was fuelled even more when genetics came in to show that variations of the OXTR gene that encodes the receptor of oxytocin were linked to deficits in social behaviour. This was shown directly in animal models, where this gene can be manipulated, but it was also found that variations in this gene could increase the risk of autism.

This closed the loop. Impaired social interaction is a diagnostic hallmark of Autism Spectrum Disorders. Autism has a very high prevalence (estimated to 1 in every 68 individuals) and there is currently no satisfactory (or even close to satisfactory) medical treatment, despite the enormous effort that has been devoted to it. In the light of this, oxytocin offered huge promise to a population that was avid for solutions. The first studies showed that, as with normal adults, single doses of oxytocin could increase social cognition of autistic children. But numbers matter. The effect was quite modest: children's performance in a task in which they had to infer the emotion of another person by looking at their eyes improved, on average, by from 45 to 49 per cent. This is still very far from people without autism, who on average perform in the same task at above 70 per cent.

Oxytocin worked, but the effect was very small, almost negligible. And there were more important reasons to temper the oxytocin hype. Most drugs behave very differently when used in a single dose compared to extended treatments of multiple doses. And, here, results from animal research were not so promising. The same drug that after being administered boosted social behaviour in mice, sheep and voles resulted in very weird social behaviour in the long term, especially after repeated exposures. And, indeed, ten years of research have not shown any consistent effect of sustained treatments of oxytocin to improve social deficits in autistic children. Adam Guastella, one of the world leaders in oxytocin investigation, published in 2016 a review paper which analyses all current evidence to conclude that repeated doses of oxytocin have very limited therapeutic potential.

Let's put all this together, because that gives us an important lesson not only on oxytocin and social cognition, but more generally on how naive interpretations of neuroscientific findings may be strongly misleading. It is true that oxytocin plays a role in social behaviour – there is ample evidence to support this idea. Oxytocin is expressed in nature during motherhood (a moment of maximal social bond), removing oxytocin usually leads to different forms of social neglect and lack of trust, and, conversely, providing oxytocin leads to increased trust, empathy, emotional recognition and understanding . . . Hacking the genes of oxytocin receptors leads to animals with very weird social patterns, and people who have atypical variations of these genes are more likely to express autism or other diseases which affect social behaviour. So evidence spans genes, molecules, pathology, and laboratory psychological experiments on humans and animals in a consistent picture. However, the fact that a molecule plays a role in this process does not imply that simply consistently boosting this molecule will lead to increasing this behaviour. However, this is often hidden or unsaid in the broad media reports of these studies, mainly following the natural desire to make a story simple, often more beautiful and optimistic than it really is.

Oxytocin is a biological and chemical trail that lays the foundations which predispose a person to cooperate, but it is a huge and unfounded leap to assume that this implies that a network of trust, love and social understanding can be built by popping a pill.

The seeds of corruption

Trust is the foundation of human society. On every scale, in every stratum, trust is the glue of institutions. It is the key to friendship and love, and the basis of commerce and politics. When there is no trust, the bridges connecting people break, and societies fall apart.

Everything collapses. And this idea of everything breaking apart is expressed in Latin as *con* (everything) and *rumpere* (to break), which is where we get the word *corrupt*. Corruption leaves nothing intact.* It destroys the fabric of society.

The world map of corruption is not hard to imagine.** The Nordic countries, Canada and Australia, are pale yellow, indicating a very low perceived level of corruption. Europe shows a gradient with corruption increasing from north to south and from west to east. The United States and Japan have intermediate orange values, and Russia and most of Asia, Africa and South America (with the notable exception of Chile and Uruguay) show up as the places in the world with the highest level of corruption.

Many economists think that endogenous, structural corruption, which is filtered through all the pores of a society, is a fundamental obstacle to development. Therefore understanding why there are very different corruption perception values is pertinent, especially when analysing how this mechanism can offer clues that might eventually change the course of things.

Rafael Di Tella, an economist, Argentinian Olympic fencer and Harvard professor, developed – along with his doctoral student Ricardo Pérez-Truglia – a modest project within this larger objective, which sought to detect one of the seeds of corruption. Rafael's premise begins with a quote from Molière: 'He who wants to kill his dog, accuses it of rabies.' What he implies is that the perpetrator of a wrong action gets away with it by assigning the blame to the victim.

From a normative perspective we should construct opinions about others based on what they have or have not done; yet we do so based on the shape of their face, the structure of their speech or the way

---

* *Con-*, as a prefix, means 'together'. So corruption is several people breaking. You cannot be corrupt all by yourself.

** In this case I am referring to the map developed by Transparency International. It doesn't directly measure corruption, but rather its perception in each society.

they walk. The consequences of Molière's conjecture are even more unsettling. It implies that we construct unfounded opinions about others to justify our aggression.

Rafael took Molière's idea to the laboratory with an ingenious experiment called the 'Corruption Game'.

Like all games derived from the 'Dictator Game', it begins with a player – the agent – who decides how to allocate twenty tokens, which were the payment for a boring job that was done by the agent and another player (the allocator), who never meet throughout the experiment. The fundamental aspects of the Corruption Game are as follows:

Some agents can choose, completely freely, how many tokens they want to keep. Others have a small margin of action: they can only choose to keep ten, eleven or twelve tokens. According to the rules of this version of the game they are forced to distribute at least eight to the other player. This controls how much the agent can mistreat the other player, so as to see later what the agent thinks about the recipient.

The recipient receives the tokens in sealed envelopes, without knowing how they were distributed, then can trade those tokens and the agent's tokens for cash. When doing so the recipient has to make a decision: trade them fairly – five dollars for each token – or trade them corruptly according to the arrangement offered to the cashier, who will pay $2.50 for each token but in exchange will offer a bribe. So the arrangement benefits the cashier and the recipient, and cheats the agent.

In this game, the agent can act in a generous or selfish way, and the recipient can act in an honest or corrupt way. The question (Molière's) is whether the selfish agents justify their action by arguing that their recipients were going to be corrupt. The fundamental key is that the tokens are in a sealed envelope and therefore, when deciding to trade them, the recipient still doesn't know how they were distributed. In this game, the player who acts corruptly does so on the basis only of personal predisposition, not for revenge or payback.

Despite that, Molière governs the game. Those agents who were offered more freedom to play aggressively tend to deem the recipients as more corrupt. And this is true both in terms of their fellow players – whom they haven't met – and their view of the general public. When we can choose to be more hostile and aggressive, we tend to think that others are corrupt. Then, all dogs are rabid.

It remained to be seen how this plot is perpetuated; how opinions emanating from our own actions can, in turn, condition what we do, leading to a domino effect of corruption in the social network. In order to find this out, Andrés Babino, a doctoral student in my lab, and I joined the team Rafael Di Tella had put together.

The key was observing how the agent acted according to the opinion they had of the other player. We created a new experiment in which the recipient had to act according to one of these three instructions: The recipient:

(1) is forced to trade each token for its face value;
(2) can choose to act corruptly or not;
(3) is forced to trade the tokens for half of their value and keep the commission – in other words, is forced to act corruptly.

It could be expected that the agent – who knew which of the three rules the other was bound by – would distribute more when knowing that the recipient would not act corruptly, slightly less when uncertain whether the other player would act corruptly, and even less when warned that the recipient was forced to act corruptly.

However, that was not what happened. In fact, the agent distributed tokens with equal generosity when knowing that the recipient had no freedom of choice. It didn't matter whether the way the recipient traded turned out to be more or less favourable for the agent. And the agent was much less generous when unsure as to what the recipient would do. In the game of beliefs and trust, it is ambiguity that's the real killer.

The same argument can be applied in the opposite way. We are hostile with those we believe could betray us. It is the fear of being made a fool of, of trusting someone who will not reward us in the same way. So, putting together the two pieces of the puzzle, our own selfish actions turn into harmful beliefs about others ('everyone is corrupt'), and ambiguity about others' beliefs ('they may be corrupt') makes us selfish and aggressive. It is a vicious circle that is only remedied by firmly sowing certainty and trust. And this, at least in the laboratory, is possible. In order to do so we must enter the deepest recesses of words and the deepest structures of the brain.

The persistence of social trust

When players make a confident, cooperative and altruistic decision in the trust game, the regions of their brain that codify dopaminergic circuits of pleasure and reward are activated. In other words, our brains react similarly when exposed to something pleasurable – sex, chocolate, money – as when displaying solidarity. Being good has value. And that explains why in all the economic games we rarely see decisions that exclusively maximize financial profit and ignore all social considerations. It turns out that there is a foundation for this hypothesis. Social capital is not only lovely and honourable – it pays.

When playing the trust game repeatedly, players learn and align in a pattern: if one player distributes generously, the other becomes progressively more generous. And the opposite is also true, if one is not generous, the other distributes in an increasingly more selfish manner. In general, the game comes to two solutions; the perfectly cooperative, in which all players win more, and the selfish, in which the first player wins less and the second gets nothing. The brain discovers the other player's inclinations using the same learning mechanism that explains the neuroscience of optimism. A person, before playing, already has an expectation of their fellow player, whether that player will cooperate or not. When they find a discrepancy, the brain's caudate nucleus activates and releases dopamine.

This produces a signal of *prediction error* that in turn makes us learn to calculate more precisely whether the other player will cooperate in the future. As this calculation becomes more exact, we learn to know our neighbours. As such there is less of a discrepancy between what is expected and what is found, and the dopaminergic signal lessens. It is the neuronal circuit of social reputation.

What's most interesting here is understanding how this slow cooking of trust thickens into an obstinate tendency to trust others. This can perhaps explain in part the idiosyncratic differences between Argentinians, Chileans, Venezuelans and Uruguayans in their predisposition to trust others or, alternatively, to become corrupt.

The key experiment was done by a neurobiologist, Elizabeth Phelps, in New York. A person repeatedly plays trust games with different players. Each of them had previously been described in a brief made-up biography that marked them as morally upstanding or immoral.

And she discovered something extraordinary in the brains of those playing with someone described as morally upstanding who nonetheless behaves selfishly. Since the brain learns from discrepancies, we would expect that a prediction error would be produced in the caudate nucleus, releasing dopamine and in turn allowing for a revision in the opinion of the other person. Their good reputation would have to be adjusted to take into account the bad action just observed. But that doesn't happen. The brain turns a blind eye when there is a discrepancy between your moral expectation of a person and their actions. The caudate nucleus does not activate, the dopamine circuits shut off and there is no learning. This obstinacy is a lasting social capital that can resist certain setbacks. Those who establish (based on the biography they were given) that the other player will act morally do not change that belief merely because they find an exception. Which is to say, the trust network is robust and sturdy. The seed of social confidence is optimism's first cousin.

We can recognize this in a more mundane situation; for example, when someone whose taste in movies we respect enthusiastically recommends a film that we think is without merit. We curse their name in the moment, but our trust in them persists. There would have to be many more failed recommendations before we began to question their judgement in recommending films to see. Yet if a person we barely know recommends a bad book to us, we will be unlikely to take their advice ever again.

To sum up . . .

Over the course of this chapter we have travelled far and wide through human decision-making, from our simplest choices to our most sophisticated and profound. The decisions that define our morality, our notion of what is fair, whom we love. Those ones that José Saramago says 'make us'.

Over the course of this journey a latent and implicit tension naturally appeared. On one hand, we spoke of the existence of a common neuronal circuit that mediates practically every human decision. On the other, we have shown that our ways of deciding are markedly personal, and that our decisions define us. Some are utilitarian and pragmatic; some are trusting and willing to take risks; others are prudent and spineless. What's more, this mishmash of decisions coexists deep within each of us. How is it possible that one single cerebral mechanism can produce such a wide range of decisions? The key is that the machine has various screws, and the way they are tightened can result in decisions that seem very different despite having structural similarities. So a slight change in the balance between the lateral frontal cortex and the medial frontal cortex defines us as cold and calculating, or emotional and hypersensitive. Often what we perceive as opposing decisions are, in actuality, only a very slight disturbance in a single mechanism.

This is not only true of the decision-making machinery. It turns out to be perhaps the essence of the biology that defines us: diversity within regularity. Noam Chomsky caused a stir when he explained that all languages, each with its own history, idiosyncrasies, usage and customs, share a common skeleton. This holds true for the language of genetics. We all roughly share the same genes; otherwise it would be impossible to talk about a 'human genome'. But the genes are not identical. For example, there are certain places on the genome – called polymorphisms – that are wildly varied and, to a large extent, define the unique individuals we each are.

Of course, these seeds take shape within a social and cultural breeding ground. Despite a genetic predisposition to and a biological seed for cooperation, it would be absurd from every standpoint to believe that Norwegians are less corrupt than Argentinians because of a different biological makeup. However, here there is an important nuance. It is not impossible – in fact, it's quite likely – that the brain's shape and organization develop differently depending on whether it's reared in a culture based on trust or distrust. It is within cultures that the machine's screws are tightened, its parameters are configured, and the results are expressed in how we make decisions and how we trust. In other words, culture and brain are intertwined in an eternal golden braid.*

---

* My humble homage to the famous book *Gödel, Escher, Bach: An Eternal Golden Braid*, by Douglas Hofstadter (published in the USA in 1979 by Basic Books, and in the UK by Penguin Books in 1980), that inspired a generation of scientists – myself included – to take the leap from more analytical and quantitative disciplines into the adventure of the brain and human thought.

CHAPTER THREE

The machine that constructs reality

How does consciousness emerge in the brain and
how are we governed by our unconscious?

It is now possible to read and explore our thoughts by decoding patterns of brain activity. With this we can begin to establish, for example, whether a vegetative patient is aware or not. We can also explore dreams and resolve whether they actually happened as we remember them, or are just a narrative created by our brain as we woke up. Who wakes up when consciousness is awoken? What happens in that precise moment?

Consciousness, like time and space, is something we all are familiar with but have trouble defining. We feel it and we sense it in others, but it is almost impossible to define what it is made of. It is so elusive that many of us often fall into different forms of dualism, evoking a non-physical and non-spatial entity to represent the conscious mind.

Lavoisier, the heat of consciousness

On 8 May 1794, in Paris, one of the finest of French scientists was guillotined by Maximilien Robespierre's troops after being accused of treason. Antoine Lavoisier was fifty years old and, among his many other legacies, left behind his *Elementary Treatise on Chemistry*, which was destined to change the world's economic and social order.

In the splendour of the Industrial Revolution, the steam engine was the motor of economic progress. The physics of heat, which up until then had been merely a matter of intellectual curiosity, took centre stage. The entrepreneurs of the age were urged to improve the efficiency of steam machines. Building on Lavoisier's studies, Nicolas Léonard Sadi Carnot, in his Borgesian *Reflections on the Motive Power of Fire*, then sketched out once and for all the ideal machine.

Seen today with privileged hindsight, there is something odd in this scientific epic that is reminiscent of the present situation with consciousness. Lavoisier and Carnot didn't have the faintest idea what heat was. Even worse, they were stuck between myths and wrong-headed concepts. For example, they believed that heat was a fluid called *caloric* that flowed from a hotter body to a cooler one. Today we know that heat is actually a state – agitated and in movement – of matter. For those versed in the subject, the idea of the caloric seems childish, almost absurd.

What will future experts in consciousness think of our contemporary ideas? Today's neuroscience is in a state of understanding equivalent to somewhere between Lavoisier and Carnot. The steam machine changed the eighteenth-century world in the same way computers and 'thinking machines' are changing ours now. Will these new machines be able to feel? Will they have their own wills, conceptions, desires and goals? Will they have consciousness? As was the case in the eighteenth century with heat, science must provide rapid responses to the understanding of consciousness, about whose fundamental substratum we still know almost nothing.

Pyschology in the prehistory of neuroscience

I like to think of Sigmund Freud as the Lavoisier of consciousness. Freud's great speculation was that conscious thought is just the tip of the iceberg, that the human mind is built on a foundation of unconscious thought. We only access consciously the conclusions, the outcomes, the actions evoked by this massively parallel device of unconscious thought. Freud made this discovery blindly, by observing remote and indirect traces of consciousness. Today, unconscious cerebral processes can be seen, brought to light in real time and with high resolution.

The bulk of Freud's work and almost all of his intellectual lineage were built on a psychological framework. However, over the course of his life, he also formed a neurophysiological theory of mental processes. This progression seems reasonable. To understand breathing, a pulmonologist analyses how the bronchioles work and why they become inflamed. In much the same way, the observation of the structure and functioning of the brain and its tangle of neurons is a natural path for those wanting to understand thought. Sigmund Freud, a brilliant professor of neuropathology in addition to his work as the founding father of psychoanalysis, declared his intentions in one of his first texts, *Project for a Scientific Psychology*, which was published posthumously: *to build a psychology that was a natural science, explaining the psychic processes as quantitative states determined by distinguishable materials of the nervous system.* He added that the particles which make up psychic matter are neurons. This last conjecture – which has rarely been recognized – reveals Freud's magnificent intuition.

In the last years of the nineteenth century, the scientists Santiago Ramón y Cajal and Camillo Golgi were embroiled in a very heated argument. Cajal maintained that the brain was made up of interconnected neurons. Golgi, on the other hand, believed that the brain was like a reticulum, like a continuous net. This epic scientific battle was settled by the microscope. Golgi, the great experimenter, developed a staining technique – still known today as *Golgi's method* – to see what was previously invisible. This stain added contrast to the grey edges on a grey background of brain tissue and made them visible in the microscope, shiny as gold. Cajal used the same tool. But he was wonderfully skilled at drawing, which made him highly observant and, where Golgi saw a continuum, Cajal saw the opposite: separate pieces (neurons) that scarcely touched. Altogether demolishing the image of science as a world of objective truths, the two bitter enemies together won the first Nobel Prize for Physiology. It is one of the loveliest examples of science celebrating, with its highest award and at the same ceremony, two opposing ideas.

Many years have passed since then, with many far more powerful microscopes, and we now know that Cajal was right. His work was the foundation of neuroscience, the science that studies neurons and the organ those neurons make up, along with the ideas, dreams, words, desires, decisions, yearnings and memories that they manufacture. But when Freud began his *Project for a Scientific Psychology* and sketched his brain model of a network of connected neurons, the debate between neurons and reticula was still unresolved.

Freud understood that the conditions were not yet ripe for a natural science of thought and, as such, he would not be the one to promote his *Project*. Yet today we – the heirs to Freud's work – are no longer working blindly as he was then, and we can take up the baton. It may now be prime time for the *Project* of conceiving a psychology based on the biology of the brain.

Freud working in the dark

In his *Project*, Freud sketched out the first neuronal network in the history of science. This network captured the essence of the more sophisticated models that today emulate the cerebral architecture of consciousness. It was made up of three types of neurons, *phi*, *psi* and *omega*, that functioned like a hydraulic device.

The *phi* (Φ) are sensory neurons and form rigid circuits that produce stereotypical reactions, such as reflexes. Freud predicted a property of these neurons that today has been proven by much experimental evidence: they live in the present. The Φ neurons fire rapidly because they are composed of permeable walls that release pressure soon after acquiring it. Thus they encode the stimulus received and, almost instantly, forget it. Freud was wrong about the physics – the neurons fire electrically and not hydraulically – but the principle is almost equivalent; the sensory neurons of the primary visual cortex are biophysically characterized by having rapid charge and discharge times.

The Φ neurons also detect our inner world. For example, they react when the body registers that hydration is necessary, by producing a feeling of thirst. So these neurons transmit an objective, a sort of *raison d'être* – searching out water in this case – but they do not have memory or consciousness.

Freud then introduced another type of neuron, called the *psi* (Ψ), which is capable of forming memories, allowing the network to detach from the immediacy of the present. Ψ neurons are made up of an impermeable wall that accumulates and stores, in isolation, our history of sensations. Today we know that the neurons in the parietal and frontal cortices – that codify working memory (active, for example, when remembering a phone number or an address for several seconds) – function similarly to Freud's conjecture. Except that, instead of having an impermeable casing, they manage to keep their activity alive through a feedback mechanism; like a loop that allows them to recoup the charge they are constantly losing. Yet long-term memories – for example, a childhood memory – work in a very different way from what Freud put forth. The mechanism is complex but, in large part, the memory establishes itself in the pattern of connection between neurons and in their structural changes, not in their dynamic electrical charge. This results in much more stable and less costly memory systems.

Freud was visionary in his anticipation of another conundrum. Since consciousness feeds on past experiences and representations of the future, it cannot be attached to the Φ system, which only codifies the present. And since the contents of consciousness – which is to say, what we are thinking – are constantly changing, they cannot correspond to the Ψ system, which doesn't change over time. With manifest annoyance, Freud then described a new system of neurons that he called *omega* (Ω). These neurons can – like those of memory

– accumulate charge over time and, therefore, organize themselves in episodes. His hypothesis was that the activation of these neurons was related to awareness and that they could integrate information over time and jump, like in hopscotch, between states to the rhythm of an internal clock.

We will see that this clock does indeed exist inside our brains, organizing conscious perception into a sequence of film stills. As we will see at the end of this chapter, the existence of such a clock can explain an intriguing and common illusion that Freud could not have seen: for example, why, when we are watching a motor race, do the wheels sometimes seem to turn in the wrong direction?

Free will gets up off the couch

One of the most powerful ideas in Freud's neuronal circuit was barely hinted at in his *Project*. The Φ neurons (sensations) activate the Ψ neurons (memory), which in turn activate the Ω neurons (awareness). In other words, consciousness originates in the unconscious circuits, not in the conscious ones. This flow set a precedent for three inter-woven ideas that proved decisive in the study of awareness:

(1) Almost all mental activity is unconscious.

(2) The unconscious is the true motor of our actions.

(3) The conscious mind inherits and, to a certain extent, *takes charge* of those sparks from the unconscious. Consciousness, thus, is not the genuine author of our (conscious) actions. But it, at least, has the ability to edit, modify and censure them.

This triad, a century later, has become tangible through experi-ments that hack into the brain, questioning and delineating the notion of free will. When we choose something, was there ever really any other option? Or was everything already determined and we only had the illusion of being in control?

Free will leaped into the scientific arena in the early 1980s with a foundational experiment by Benjamin Libet. The first trick was to reduce freedom of expression to its most rudimentary form: a person freely choosing when to push a button. That relegated it to a single act of just one bit. It is a simple, minimal freedom, but freedom nonetheless. After all, we are all free to push the button when we feel like it. Isn't that so?

Libet understood that in order to reveal this fundamental enigma he had to register three channels simultaneously.

First of all, the exact moment in which a supposedly free decision-maker believes he or she is making a decision. Imagine, for example, that you are on a diving board, deliberating over whether or not to dive into a pool. The process can be long, but there is a fairly precise moment in which you decide whether to dive. With a high-precision watch, and switching the vertigo of the diving board for a mere button, Libet recorded the exact moment in which the participants felt they were making the decision to push the button. This measurement reflects a sub-jective belief, the story that we tell ourselves about our own free will.

Libet also recorded participants' muscular activity in order to pinpoint the precise moment when they made use of their supposed freedom and pushed the button. And he discovered that there was a small lag of one third of a second between when they believed they had made a decision and when they carried it out. This is reasonable and simply reflects the conduction speed of the motor signal needed to execute the action. To measure brain activity, he used an electroencephalogram (EEG); a few small electrodes placed at the surface of the scalp. And the extra-ordinary finding in Libet's experiment showed up in this third channel. He discovered a trail of cerebral activity that allowed him to identify the moment in which participants would press the button, half a second before they themselves recognized their intention. It was the first clear demonstration in the history of science of an observer able to codify cerebral activity in order to predict another person's intention. In other words, to read their thoughts.

Libet's experiment gave rise to a field of investigation that produced countless new questions, details and objections. Here we will only look at three of them. The first two are easily solved. The third opens up a door to something about which we have very little knowledge.

A general criticism of this experiment (made by Libet himself and many other scientists who followed this work) is that the moment in which the decision is made is not always clear. And even if it were, his method allowed for a degree of imprecision in the recording. A second natural objection is that before making a decision there is a process of preparation. One can get into diving position before having decided to dive into the pool. Many of us, in fact, glumly retreat from the board without taking the plunge. Perhaps what Libet observed was the brain's preparatory circling around the decision.

These two objections are resolved in a contemporary version of Libet's experiment, conducted by John-Dylan Haynes in 2008, with two subtle but decisive differences. First of all, the resolution of the measuring instrument is improved by using magnetic resonance instead of the electroencephalogram with fewer channels than Libet employed, allowing for greater precision in decoding cerebral states.

Secondly, participants' freedom of expression is doubled: they can now choose between two buttons. This minimal variation allowed Haynes to distinguish the choice (right or left button) from the action (the moment of pushing one of the buttons).

With this addition of the second button and the new technology, the magnifying glass used for searching out an unconscious seed in our apparently free and conscious decision-making became much more effective. Based on the pattern of activity in a region of the frontal cortex, it was possible to decipher the content of a decision ten seconds before a person *felt* that they were making it. The region of the brain that denotes our future actions is vast but specifically includes a zone in the more frontal and medial part that we are already familiar with: the Brodmann Area 10, which coordinates inner states cohesively with the outer world. In other words, when a person actually makes a decision, they do not know that in fact it had already been made a few seconds earlier.

The more difficult problem with Libet's experiment is knowing what happens if someone intentionally decides to push the button but then deliberately halts before doing so. Libet himself responded to this, arguing that consciousness has no vote but does have a veto. Which is to say, it doesn't have the capability or the freedom to set an action into motion – the task of the unconscious – but it can, once this action becomes observable to it, manipulate it and eventually stop it. Consciousness, in this scenario, is like a sort of *preview* of our actions in order to filter and mould them.

In Libet's experiment, if someone decides to press the button and then changes their mind, a series of cerebral processes can be observed; the first codifies the intent to act that is never realized; later, a very different second process reveals a system of monitoring and censorship governed by another structure in the frontal part of the brain that we have already looked at, the anterior cingulate.

Does the conscious decision to halt an action also stem from another unconscious seed? This is still – as I understand it – a mystery. The problem is sketched in Borges's fable about chess pieces:

> God moves the player, who moves the piece.
> What God behind God gives rise to the plot
> of dust and time and dreams and agony?

In this endless recursion of wills that control wills (the decision to dive into the pool, then the hesitation and the decision to stop, then another that soothes the fear so the first decision can continue its course . . .) a loop emerges. It is the brain's ability to observe itself. And this loop is perhaps, as we will see further on, the basis of the principle of consciousness.

The interpreter of consciousness

The brain's two hemispheres are connected by a massive structure of neuronal fibres called the corpus callosum. It is like a system of bridges that coordinates traffic between the two halves of a city divided by a river; without the bridges, the city is split in two. Without the corpus callosum, the cerebral hemispheres are isolated from each other. Some years back, in order to remedy some forms of epilepsy that were resistant to pharmaceutical treatment, some patients underwent a corpus callosotomy, a surgical procedure in which the two hemispheres were split apart. Epilepsy is, to a certain extent, a problem of brain connectivity that results in cycles of neuronal activity that feed on themselves. This surgical procedure interrupts the flow of currents in the brain and is a dramatic but effective way of putting paid to these cycles and, with them, epilepsy.

What happens to the language, emotions and decisions of a body governed by two hemispheres that no longer communicate with each other? The methodical answer to this question, which also allows us to understand how the hemispheres distribute functions, earned Roger Sperry the Nobel Prize – shared with Torsten Wiesel and David Hubel – in 1981. Sperry, along with his student Michael Gazzaniga, discovered an extraordinary fact that, just like Libet's experiment, changed how we understand our construction of reality and, with it, the fuel of consciousness.

Without the corpus callosum, the information available to one hemisphere cannot be accessed by the other. Therefore, each hemisphere creates its own narrative. But these two versions are enacted by the same body. The right hemisphere only *sees* the left part of the world and also controls the left part of the body. And vice versa. Additionally, a few cognitive functions are fairly compartmentalized in each hemisphere. Typical cases are language (left hemisphere) and the ability to draw and represent an object in space (right hemisphere). So if patients with separated hemispheres are shown an object on the left side of their visual field, they can draw it but not name it. Conversely, an object to the right of their visual field is accessed by the left hemisphere and as such can be named but not really drawn.

Sperry's great discovery was understanding how our consciousness creates a narrative. Imagine the following situation: patients with separated hemispheres are given an instruction in their left visual field; for example, that they will be paid money to lift up a bottle of water. Since it was presented to the left visual field, this instruction is only accessible to the right hemisphere. The patients pick up the bottle. Then they ask the patients' other hemisphere why they picked it up. What do they respond? The correct answer, from the perspective of the left hemisphere – which did not see the instruction – should be 'I don't know.' But that's not what the patients say. Instead, they invent a story. They put forth reasons, such as that they were thirsty or because they wanted to pour water for someone else.

The left hemisphere reconstructs a plausible story to justify the participants' action, since the real motive behind it is inaccessible to them. So the conscious mind acts not only as a front man but also as an interpreter, a narrator who creates a story to explain in hindsight our often inexplicable actions.

'Performiments': freedom of expression

Perhaps the most striking aspect of these fictional narratives created by patients with separated hemispheres is that they aren't deliberate falsifications to hide their ignorance. The narrative is true, even to those fabricating it. Consciousness's ability to act as an interpreter and invent reasons is much more common than we recognize.

A group of Swedes from Lund – near Ystad, where the detective Kurt Wallander also deals, in his own way, with the intricate tricks of the mind – produced a more spectacular version of the interpreter experiment. In addition to being scientists, these Swedes are magicians and, as such, know better than anyone how to force their audience's choices, in making them believe illusions in a magic show as well as making them think they've made a decision completely freely in a science laboratory. Their way of putting free will in check is the show business equivalent of the project begun by Libet.

The experiment or trick, which here is the same thing, works this way: people are shown two cards, each showing a woman's face, and they must choose which woman they consider more attractive and then justify their decision. That much is pretty straightforward. But sometimes the scientist – who also acts as the magician – gives the participants – who also act as the audience – the card they didn't choose. Of course, the scientist does so using sleight of hand so that the switch is imperceptible. And then something extraordinary happens. Instead of saying, 'Excuse me, I chose the other card', most of the participants start to give arguments in favour of a choice they actually never made. Again they resort to fiction; again our interpreters create a story in retrospect to explain the unknown course of events.

In Buenos Aires, I set up, with my friend and colleague Andrés Rieznik, a combination of magic and research to develop our own *performiments*, performances that are also experiments. Andrés and I were investigating *psychological forcing*, a fundamental concept in magic that is almost the opposite of free will. It uses a set of precise tools to make spectators choose to see or do what the magician wants them to. In his book *Freedom of Expression*, the great Spanish magician Dani DaOrtiz explains exactly how the use of language, pacing and gaze allows magicians to make audience members do what they want them to. In the performiments, when the magician asks the crowd whether they saw something or not, or whether they chose the card 'they really want', the performer is actually following a precise, methodical script to investigate how we perceive, remember, and make decisions.

Using these tools we proved what magicians already knew in their bones: spectators don't have the slightest idea that they are being forced and, in fact, believe they are making their choices with complete freedom. The spectators later create narratives – sometimes very odd ones – to explain and justify choices they never made, but truly believe that they have made.

We then moved this experiment from the stage to the lab. There we performed an electronic version of a forcing magic trick. We showed participants a very rapid sequence of cards. One of them was presented for slightly more time. This change remained unnoticed to our participants but made them choose in almost half of the trials the 'forced' card. The advantage of doing this experiment in the laboratory is that we could measure, while participants were observing the flashing deck and making their choices, their pupil dilation, an autonomic and unconscious response that reflects, among other things, a person's degree of attention and concentration. And with this we discovered that there are indications in the spectators' bodies that reveal whether the choice was freely made or not. Approximately one second after a choice, the pupil dilates almost four times more when people choose the forced card. In other words, the body knows whether it has been forced to choose or not. But the spectator has no conscious record of that. So our eyes are more reliable indicators of the true reasons behind a decision than our thoughts.

These experiments deal with the old philosophical dilemma of responsibility and, to a certain extent, question the simplistic notion of free will. But they do not topple this notion, not by a long chalk. We don't know where or how Libet's unconscious spark originates. At this point we can only make conjectures about the answers to these questions, as Lavoisier did with his theory of the caloric.

The prelude to consciousness

We saw that the brain is capable of observing and monitoring its own processes in order to control them, inhibit them, shape them, halt them or simply manage them, and this gives rise to a loop that is the prelude to consciousness. Now we will see how three seemingly innocuous and mundane questions can help us to reveal and understand the origin of and reason for this loop, and its consequences.

Why can't we tickle ourselves?

We can touch ourselves, watch ourselves, caress ourselves, but we can't tickle ourselves. Charles Darwin, the great naturalist and father of contemporary biology, took on this question in depth and with rigour. His idea was that tickling only works if one is taken by surprise, and that unexpected factor disappears when we do it to ourselves. It sounds logical, but it is false. Anyone who has ever tickled someone knows that it is just as effective – or even more so – if the victim is warned ahead of time. The problem of the reflexive impossibility of tickling oneself then becomes much more mysterious; it is not only that it isn't a surprise.

In 1971, Larry Weiskrantz published an article in *Nature* entitled 'Preliminary Observations on Tickling Oneself'. For the first time, tickles took centre stage in the research on consciousness. Then it was Chris Frith, another illustrious figure in the history of human neuroscience, who began to take tickling seriously – despite the oxymoron – as a privileged window into the study of the conscious mind.

Frith built a tickler, a mechanical device to allow people to tickle themselves. The detail that converted the game into science is the ability to change the intensity and delay in its action. When the tickler works with a scarce half-second delay, the tickling is felt as if someone else were doing it. When some time passes between our actions and their consequences, that produces a strangeness which makes them be perceived as being performed by others.*

Why doesn't the image we're looking at move when we move our eyes?
Our eyes are in constant movement. They make an average of three *saccades* or abrupt shifts per second. In each one, our eyes move at top speed from one side of an image to the other. If our eyes are moving all the time, why is the image they construct in our brains still?

We now know that the brain edits the visual narrative. It is like the camera director of the reality we construct. The stabilization of the image depends on two mechanisms that are now being tested out in digital cameras. The first is *saccadic suppression*; the brain literally stops recording when we are moving our eyes. In other words, for the split-second when our eyes are in motion we are blind.

This can be shown in a quick experiment at home: stop in front of a mirror and direct your gaze to one eye and then the other. When you do this your eyes will, of course, move. Yet, what you will see in the mirror are your immobile eyes. That is the consequence of the microblindness that occurs in the exact moment that our eyes are moving.

Even if we edit the mental movie as our eyes move, there is still a problem. After a saccade, the image should move the way it does

---

* There are other types of strangeness that can be achieved by time manipulations. Bill Viola, in a video installation from 1955 called *The Greeting*, re-created a Mannerist painting. At first glance, it appeared to be an image of three women. When you look more carefully you realize, almost by chance, that the women are getting closer. But it all happens so slowly that it is impossible to associate the images with the movement. After ten minutes, the women are hugging. It has been said of Bill Viola that he doesn't introduce images into time, but rather time into images.

in home movies or in Dogme films, when the frame instantly shifts from one point in the image to another. But that doesn't happen. Why not? It turns out that the receptive fields in the neurons of the visual cortex – somewhat analogous to the receptors that codify each pixel of an image – also move to compensate for the eye movement. That generates a smooth perceptive flow, in which the image remains static despite the frame constantly shifting. This is one of many examples of how our sensory apparatus reconfigures drastically according to the knowledge that the brain has of the actions it is going to carry out. Which is to say, the visual system is like an active camera that knows itself and changes its way of recording depending on how it is planning to move. This is another footprint of the beginning of the *loop*. The brain informs on itself, it has a record of its own activities. This is the prelude to consciousness.

While in a very different framework, this is the same idea that governs the impossibility of tickling ourselves. The brain foresees the movement it will make, and that warning creates a sensory change. This anticipation cannot work consciously – one cannot deliberately avoid feeling tickles, nor voluntarily edit the visual flow – but therein lies the seed of consciousness.

How do we know that the voices in our head are ours?

We talk to ourselves all day long, almost always in a near whisper. In schizophrenia, this dialogue melds with reality in thoughts plagued with hallucinations. Chris Frith's thesis is that these hallucinations result from the inability of schizophrenic patients to recognize that they are the creators of their inner voices. And since they don't recognize them as their own, as with tickling, they cannot control them.

This argument withstands fierce experimental scrutiny. The region of the brain that codifies sounds – the auditory cortex – responds in a subdued way when we hear our own voice in real time. But if the same speech is played back and heard in a different context, it generates a cerebral response of greater magnitude. This difference is not observed in the auditory cortices of schizophrenics, whose brains do not distinguish when their voices are presented in real time or in a replay.

It turns out to be very difficult to understand the mind's quirks when we don't experience them ourselves. How can someone perceive their inner mental conversations as if they were external voices? They are inside us, we produce them, they are obviously ours. Yet there is a space in which almost all of us make the same mistake, again and again: in dreams. They are also fictions created by our imagination, but dreams exercise their own sovereignty; it is difficult, almost impossible, for us to appropriate their stories. What's more, many times it is impossible to recognize them as dreams or products of our imaginations. That is why we feel relief when we wake up from a nightmare. In some sense, then, dreams and schizophrenia have similarities, since they both revolve around not recognizing the authorship of our own creations.*

In short: the circle of consciousness

These three phenomena suggest a common starting point. When an action is carried out, the brain not only sends a signal to the motor

---

* An excerpt from the second night of Borges's *Seven Nights*. 'I met up with a friend, a friend I don't know: I saw him and he had changed so much. I had never seen his face but I knew that that face couldn't be his. He was very changed, very sad. His face was marked by sorrow, by illness, perhaps by guilt. He had his right hand inside his jacket (this is important in the dream). I couldn't see his hand, which was hidden over his heart. Then I hugged him, I felt that he needed my help: "But, my poor So-and-so, what happened to you? You're so changed!" He replied, "Yes, I've changed a lot." He slowly pulled out his hand. I could see that it was a bird claw. The strange thing is that the man's hand had been hidden from the very beginning. Without realizing it, I had prepared that invention: that the man had a bird claw and I could see how terrible his change was, how terrible his fate, how he was turning into a bird.'

cortex – so the eyes and hands move – but also alerts itself to readapt beforehand. In order to be able to stabilize the camera, in order to be able to recognize inner voices as its own. This mechanism is called *efferent copy*, and it is a way that the brain has of observing and monitoring itself.

We have already seen that the brain is a source of unconscious processes, some of which are expressed in motor actions. Shortly before being carried out, they become *visible* to the brain itself, which identifies them as its own. This sort of cerebral signature has consequences. It happens when we move our eyes, when we can't tickle ourselves, when we mentally recognize our own voice; we can think generically of this mechanism as an internal communication protocol.

A useful analogy here might be how, when a company decides to launch a new product, it lets its different departments know so that they can coordinate the process: marketing, sales, quality control, public relations, etc. When the company's internal communications (its efferent copy) fail, incoherencies result. For example, the purchasing group observes that there is less availability of some raw material and has to guess the reason because it is not aware of the new product launch. In the same way, due to the lack of internal information, the brain comes up with its idea of the most plausible scenario for explaining the state of things. We can see in this analogy a metaphor for schizophrenia. It serves to convey the image of how delusions arise from a deficit in an internal communication protocol.

This is of course just an exercise of thought. There is no doubt that the company is not conscious of itself. But it sets a prerequisite of consciousness when it begins to inform itself of its own knowledge and its own states in a way that can be broadcast to different sections. However, this discussion may become less rhetorical and more concrete in the near future, when we build machines that can express all the features of consciousness. Will we consider them conscious? What rights and obligations will they have?

The physiology of awareness

We live in unprecedented times, in which the factory of thoughts has lost its opacity and is observable in real time. How does brain activity change when we are conscious of a process?

The most direct way to tackle this question is to compare cerebral responses to two identical sensory stimuli that, due to internal fluctuations – in attention, concentration or waking state of the subjects – follow completely different subjective trajectories. In one case we consciously recognize the stimulus: we can talk about it and report on it. The other occurs without a conscious trace, affecting the sensory organs and continuing its cerebral trajectory in a way that doesn't result in a qualitative change in our subjective experience. This would be an unconscious or subliminal stimulus. Let's think about the most tangible and common case of an unconscious stimulus: imagine that someone is speaking to us while we are placidly falling asleep. The words progressively vanish; we still hear sound arriving to our ears.

Let us begin by seeing how a subliminal image is represented in the brain. Sensory information arrives, for example, in the form of light to the retina, and turns into electrical and chemical activity that spreads through the axons to the thalamus, in the very centre of the brain. From there, the electrical activity spreads to the primary visual cortex, located in the back of the brain, near the nape. So, about 170 milliseconds after a stimulus reaches the retina, a wave of activity occurs in the brain's visual cortex. This delay is not only due to the conduction times in the brain but also to the construction of a cerebral state that codifies the stimulus. Our brain lives, literally, in the past.

The activation of the visual cortex codifies the properties of the stimulus – colour, luminosity, movement – so well that in the laboratory an image can be reconstructed based on the pattern of cerebral activation produced. What is most surprising is that this happens even if the image is presented subliminally. In other words, an image remains recorded for a while (at least) in the brain, even though that cerebral activity doesn't produce a conscious mental image. With the proper technology, this recorded image can be reconstructed and projected. So today we are literally able to see the unconscious.

This whole river of cerebral activity that happens in the underground of consciousness is similar to that provoked by a *privileged* stimulus which is able to access the narrative of consciousness. This is interesting in and of itself and represents the cerebral trail of unconscious conditioning that was sketched out by Freud. But the unconscious is, in phenomenological and subjective terms, very different from the conscious mind. What happens in the brain to differentiate one process from the other?

The solution is very similar to what makes a fire spread or a tweet go viral. Some messages circulate in a local atmosphere, and certain fires remain confined to small sectors of a forest. But every once in a while, due to circumstances intrinsic to the object (the content of the tweet or the intensity of the fire) or the network (the dampness of the ground or the time of day in a social network), the fire and a tweet take over the entire network. They spread massively in an expansive phenomenon that begins to fuel itself. They become viral, and uncontrollable.

In the brain, when the intensity of the neuronal response to a stimulus exceeds a certain threshold, a second wave of cerebral activity is produced, about 300 milliseconds after the stimulus occurs. This second wave of activity is no longer confined to the brain regions related to the sensory nature of the stimulus (the visual cortex for an image or the auditory cortex for a sound), like a wildfire that has spread throughout the brain.

If this second massive wave takes over the brain almost entirely, the stimulus is conscious. Otherwise, it isn't. The cerebral activity leaves a mark that is a sort of digital signature of consciousness, allowing us to know if a person is conscious or not, to access their subjectivity, and to know the contents of their mind.

This wave of cerebral activity, which is only registered in conscious processes, is:

(1) MASSIVE. A state of great cerebral activity propagated and distributed throughout the entire brain.

(2) SYNCHRONIZED AND COHERENT. The brain is made up of different modules that carry out specific activities. When a stimulus accesses consciousness, all of these cerebral modules synchronize.

(3) MEDIATED. How does the brain manage to create a state of massive, coordinated activity among modules that usually work independently? What performs that task? The answer is, again, analogous to the social networks. What makes information go viral? On the web there are hubs or traffic centres that function as huge information propagators. For example, if Google prioritizes a particular piece of information in a search, its diffusion increases.

In the brain there are at least three structures that carry out that role:

(a) The frontal cortex, which acts sort of like a control tower.

(b) The parietal cortex, which has the virtue of establishing dynamic route changes between different brain modules, sort of like a railway switch that allows a train to pass from one track to another.

(c) The thalamus, which is in the centre of the brain, connected to all the cortices, and in charge of linking them all. When the thalamus is inhibited, it strongly disassociates traffic in the cerebral network – as if one day Google shut down – and the different modules of the cerebral cortex cannot synchronize themselves, making consciousness vanish.

(4) COMPLEX. The frontal cortex, the parietal cortex and the thalamus allow the different actors within the brain to act in a coherent manner. But how coherent does the activity in the brain have to be in order for it to be effective? If the activity were completely disorganized, the traffic and flow of information between different modules would become

impossible. Full synchronicity, on the other hand, is a state in which ranks and hierarchies are lost, and where modules and compartments that can realize specialized functions are not formed. In the extreme states of completely ordered or chaotic cerebral activity, consciousness disappears.

This means that the synchronization must have an intermediate degree of complexity and internal structure. We can understand it with an analogy to musical improvisation; if it is totally disorganized, the result is pure noise; if the music is homogeneous and no instrument offers any variation from the others, all musical richness is lost. What's most interesting happens at an intermediate degree between those two states, in which there is coherence between the different instruments but also a certain freedom. It is the same with consciousness.

Reading consciousness

In July 2005 a woman had a car accident that left her in a coma. After the routine procedures, including surgery to reduce pressure in the brain caused by various haemorrhages, the days passed with no signs of her recovering consciousness. From that moment on, and over weeks and months, the woman opened her eyes spontaneously, and had cycles of sleep and wakefulness and some reflexes. But she made no gesture that indicated a voluntary response. All these observations corresponded with the diagnosis of a vegetative state. Was it possible that, against all clinical evidence, the patient had a rich mental life, with a subjective landscape similar to that of a person in a state of full consciousness? How could we know? How can we investigate the mental life in someone else's mind if they can't communicate their thoughts?

In general, other people's mental states – happiness, desire, boredom, weariness, nostalgia – are inferred by their gestures and their verbal expression. Language allows us to share, in a more or less rudimentary way, our own private states: love, desire, pain, a special memory or image. But if you are unable to externalize this mental life, as happens for example while sleeping, a person is locked in. Vegetative patients do not externalize their thoughts and, therefore, it was normal to assume that they might not have consciousness. This has all changed. The properties of conscious activity that we enumerated become dramatically relevant because they allow us to decide, in an objective way, whether a person has signatures of consciousness. They work as a tool to read and decipher other people's mental states, something that becomes more pertinent when it is the only way to do so, such as in the case of vegetative patients.*

Observing the imagination

Some seven months after the car accident that left her in a vegetative state, doctors made a study of the woman using functional magnetic resonance imaging. Can tracing cerebral activity provide a view of her thoughts? Her brain activity, when hearing different phrases, was comparable with that of a healthy person. Most interestingly, the response was more pronounced when the phrase was ambiguous. This suggested that her brain was struggling with that ambiguity, which indicated an elaborate form of thinking. Perhaps the woman wasn't truly in a vegetative state? The observations of her brain were not enough to respond conclusively to such a significant question. During deep sleep or under anaesthesia – where one presumes that a person is indeed unconscious – the brain also responds elaborately to phrases

---

* Naming is an art; sometimes a terrible art. The term *vegetative* is already revealing: it presupposes an organism that carries out its life cycle without being a true protagonist in its own actions. The organism has a metabolism, regulates its vital functions and even has some automatic emotional responses, but nothing that is represented by an agent – a being – who controls the life of its own mind and body.

and sounds. How can the signature of consciousness be more precisely examined?

When a conscious person imagines that they are playing tennis, the part of the brain that activates the most is known as the supplementary motor area (SMA). This region controls muscular movement.* On the other hand, when someone imagines walking through their house – we can all mentally follow a route through many maps, train lines, friends' grandmothers' houses, cities, trails – a network activates, which primarily involves the parahippocampus and the parietal cortex.

The regions that activate when someone imagines they are playing tennis are very different from the ones that activate when they imagine walking through their house. This can be used to decipher thought in a rudimentary but effective way. It is no longer necessary to ask someone if they are imagining tennis or imagining moving through their house. It is possible to decode it precisely just by observing their cerebral activity. In effect, we can read someone else's mind; at least along a binary code of tennis or house. This tool becomes particularly relevant when we cannot ask questions. Or, more accurately, when the person cannot answer them.

Can that 23-year-old vegetative woman imagine? The British neuroscientist Adrian Owen and his colleagues posed that question in the resonator in January 2006. They asked the patient to imagine tennis and then imagine walking around her house, then tennis again, then walking again, and so on, alternately imagining one activity and then the other.

The cerebral activation was indistinguishable from that of a healthy person. So it can be reasonably inferred that she was

---

* This should not be misinterpreted to mean that this is the brain's *tennis* region. No such thing exists. This region carries out a function of muscular activity coordination and, of course, would also activate when imagining a dance, a dive or a game of jai alai.

capable of imagining and, therefore, that she had conscious thoughts that could not have been supposed by her doctors based on clinical observation.

The moment when she managed to break through the opaque shell that had confined her thoughts for months – as Owen and his team observed her thinking directly in her brain – was a landmark in the history of human communication.

Shades of consciousness

The demonstration with tennis and spatial navigation has an even greater significance: it is a way of communicating. A rudimentary but effective one.

With this we can establish a sort of Morse code. Every time you want to say 'yes', imagine you are playing tennis. Every time you want to say 'no', imagine you are walking through your house. In this way, Owen's group could communicate for the first time with a vegetative patient, who was twenty-nine years old. When they asked him if his father's name was Alexander, the supplementary motor area activated, indicating that he was imagining tennis and meaning, in this code, a 'yes'. Then they asked the patient if his father was named Tomás and the parahippocampus activated, indicating spatial navigation and representing a 'no' in the code they'd established. They asked him five questions, which he responded to correctly with this method. But he didn't respond to the sixth.

The researchers argued that perhaps he hadn't heard the question, or maybe he had fallen asleep. This, of course, is very difficult to determine in a vegetative patient. At the same time, the result shows the infinite potential of this window on to a previously inaccessible world, as well as raising a certain scepticism.

This last statement, as I see it, is a pertinent and necessary warning about a 'broken link' in science communication, one that distorts reality. The traces of communication in vegetative patients are promising but still very rudimentary. It is likely that the current limitations can be overcome with improving technology, but it is deceptive to believe – or make others believe – that these measures indicate an awareness that is similar in form or content to that of a normal life. Perhaps it is a much more confused and disordered state, a disintegrated, fragmented mind. How can we know?

Tristán Bekinschtein, a friend and companion in many adventures, and I set out to approach this question. Our approach was somewhat minimalist: we tried to identify the minimum behaviour that defines consciousness. And we found the solution in an experiment that Larry Squire, the great neurobiologist of memory, had done by adapting Pavlov's classic demonstration.

The experiment works like this: a person watching a film – by Charlie Chaplin – hears a sequence of tones: beep buup beep beep buup . . . One is high-pitched and the other is low. Each time the low tone* is heard, a second later that person receives a slightly annoying burst of air on to one eyelid.

Close to half of the participants recognized the structure: the low tone was always followed by the puff of air. The other half didn't learn the relationship; they didn't discover the rules of the game. They could describe the tones and the bothersome burst of air but didn't perceive any relationship between them. Only those who consciously described the rules of the game acquired the natural reflex of closing their eyelid after the low-pitched tone, anticipating the air and attenuating its bothersome effect.

---

* The *buup*, obviously, otherwise Bouba (see p. 9) wouldn't be what it is.

Squire's results seem innocent enough but are actually quite meaningful. This extremely simple procedure establishes a minimal test – a Turing test – for the existence of consciousness. It is the perfect bridge between what we wanted to know – whether vegetative patients have consciousness – and what we could measure – if they blink or not, something that vegetative patients can do – so Tristán and I built that bridge to measure consciousness in vegetative patients.

I remember the moment as one of the few in my scientific career when I felt the giddiness of discovery: Tristán and I were in Paris, and we discovered that a patient was capable of learning just as well as people with full consciousness. Then, by laboriously repeating the procedure, we found that only three out of the thirty-five patients we had examined showed this residual form of consciousness.

We spent many years refining the process in order to explore in further detail how reality is seen from the perspective of a vegetative patient who has traces of consciousness. In order to do so, Tristán adapted the experiment with beeps and puffs of air into a more sophisticated version. This time, the participants had to discover that different words in a single semantic category were preceded by a puff of air. Being fully aware wasn't enough to learn that relationship; they also had to be able to direct their attention to the words. Which is to say, those who were distracted learned in a much more rudimentary way.

So we were able to question the attention focus of vegetative patients and we found that their way of learning was very similar to that of distracted people. Perhaps this is a better metaphor for the functioning of the minds of some vegetative patients with signs of consciousness: flightier ways of thinking, in a much more fluctuating, less attentive and more disordered state.

Consciousness has many signatures. These can be naturally combined in order to determine whether a person has consciousness, but the argument for or against the determination of a patient's conscious-ness can never be definitive or certain. If their frontal and thalamic activity is normal, if their cerebral activity has an intermediate range of coherence, if certain stimuli generate synchronous activity and after about 300 milliseconds produce a massive wave of cerebral activity, and if, in addition, there is a trail of directed imagination and forms of learning that require consciousness – if all of these conditions are simultaneously found, then it is very plausible that the patient has consciousness. If only some of them exist, then there is less certainty. All these tools in conjunction are the best means we have today of coming up with an objective diagnosis of conscious activity.

Do babies have consciousness?

Research into others' thoughts is also a window into the mysterious universe of newborns' thinking. How does consciousness develop before a child can express it in gestures and concise words?*

Newborns have a much more sophisticated and abstract thought organization than we imagine. They are able to form numerical and moral concepts, as we saw in Chapter 1. But these ways of thinking could be unconscious and don't tell us much about the subjective experience during development. Are babies consciously aware of what is happening to them, of their memories, their loved ones, or their sadness? Or do they merely express reflexes and unconscious thinking?

This is a very new field of investigation. And it was my friend and colleague of many years Ghislaine Dehaene-Lambertz who took the first stab at it. The strategy is simple; it involves observing whether babies' brain activity has the cerebral signatures that indicate con-scious thought in adults. The trick is very similar to the experiment to

---

* The etymology of the word infant – the prefix *in* and *fari*, to speak – is precisely that, speechless.

understand how, in the adult brain, a conscious process diverges from an unconscious one.

At five months old, the first phase of cerebral response is practically established. This phase codifies visual stimuli, independently of whether they access consciousness. At this point, the visual cortex is already able to recognize faces and does so in similar ways and at a similar speed as adults do.

The second wave – exclusive of conscious perception – changes during development. At one year of life it is already practically consolidated and presents very similar forms to an adult's but with a revealing exception: it is much slower. Instead of at 300 milliseconds, it consolidates almost a second after seeing a face, as if babies' conscious film had a slight lag, like when we watch a game being broadcast with a delay and we hear our neighbours shouting 'goal' before we see it.

This lag in response is much more exaggerated in five-month-old babies. Long before developing use of speech, before crawling, when they can barely sit up, babies already have cerebral activity denoting an abrupt and extended response throughout the brain, which persists after the stimulus disappears.

It is the best proof we have for supposing that they have consciousness of the visual world. Surely less anchored to precise images, probably more confused, slower and hesitant, but consciousness nonetheless. Or at least that is what their brains tell us.

This is the first approximation in science to navigating a previously completely unknown territory: babies' subjective thought. Not what they are able to do, respond to, observe or remember, but something much more private and opaque, that which they are able to perceive from their conscious minds.

Deciding on the state of consciousness of a baby or a person in a vegetative state is no longer merely deliberating intuitions. Today we have tools that allow us to enter – live and direct – into the factory of thought. These tools allow us to break through one of the most hermetic and opaque barriers of solitude.

Today we still know very little about the material substratum of consciousness, as was the case before with the physics of heat. But what's most striking is that despite so much ignorance we can today manipulate consciousness: turn it on and off, read it and recognize it.

CHAPTER FOUR

Voyages of consciousness (or consciousness tripping)

What happens in the brain as we dream; is it possible for us to decipher, control and manipulate our dreams?

Altered states of consciousness

They are both lying down. He is telling her a story he's told her a thousand times, in a low, monotone voice. He pushes out the air that makes his vocal cords vibrate. The sound is modulated by his tongue, lips and palate. In less than a thousandth of a second, that wave of sound pressure bounces in his daughter's ear. The sound again becomes movement in her eardrum. That movement activates some mechanical receptors at the tip of the hair cells, a magnificent piece of biological machinery that converts the vibrations of the air into electrical pulses. Each swing of those cells opens up microscopic channels in the membranes through which ions slip in and generate a current that spreads throughout the auditory cortex, and this neuronal activity encodes the words that she, as always, repeats in a whisper. The same words that sound in her father's deep, monotone voice with delicate inflections now live in the narrative she constructs in her mind when she hears the story that she has already heard a thousand times before.

Now she breathes deeper, yawns, her body trembles briefly. She sleeps. He continues with the story, without changing the rhythm or the volume or the cadence. The sound spreads like before and strikes his daughter's eardrum, displacing hair cells and leading the ion current to activate the neurons of her auditory cortex. Everything is the same, but she is no longer creating the story in her mind. She no longer repeats the words in a whisper. Or does she? Where do the words we hear in our sleep go to?

Tristán Bekinschtein decided to take on this question, creating a simple, boring, routine experiment that was perfect for falling asleep. A reciting of words. A childhood game. It's not the typical image we have of a laboratory experiment; in fact, it takes place in a bed where someone is listening to a soporific, repetitive voice: elephant, chair, table, squirrel, ostrich . . . Each time the name of an animal is heard, they must move their right hand; if it is a piece of furniture, their left. It is easy and hypnotic. Soon the responses come intermittently. Sometimes they are extremely slow and finally they disappear. The person's breathing is deeper and the electroencephalogram shows a synchronous state, meaning that they are already asleep. The words continue, as if through inertia, like in the father's story as he presumes that his daughter is listening in her sleep.

Through observing the mark left by voices in the transitions to sleep, Tristán discovered that in the sleeping brain these voices turn into words, and those words acquire meaning. What's more, the brain continues playing the same game; the cerebral region that controls the right hand activates each time an animal is mentioned, and the region that controls the left hand activates each time a piece of furniture is mentioned, just as was established in the rules of the waking game.

Consciousness has an on/off switch. During sleep, in a coma state or under anaesthesia, the switch turns off and so does consciousness. Some cases are drastic, and consciousness turns off unambiguously.

At other times, like in the transition to sleep, it fades out gradually and intermittently. When the switch is turned on, the cerebral activity associated with states of consciousness assumes different forms; we saw, for example, that the consciousness of very young children operates on a different time scale, and that schizophrenics are unable to recognize that they are the owners of the voices in their heads, creating a distortion of the narrative.

Nocturnal elephants

We can think of dreams as fertile ground for mental simulation that does not involve the body. This disconnection between the mind and the body is literal; when we dream there is an inhibition of the motor neurons through which the brain controls and governs the muscles, generating a brain chemistry that is very distinct from our waking one.

Normally there is a synchronicity between the return to a waking state – characterized by organized conscious thought – and cerebral contact with the body. But sometimes those two processes get out of synch and we wake up without having regained chemical contact with our bodies. This is called sleep paralysis and it is experienced by between 10 and 20 per cent of the population. It can be agonizing: complete paralysis with full lucidity. Yet after a few minutes it goes away on its own, and the brain is once again in contact with the body. And the opposite can also happen, when the brain does not disconnect from the muscles during sleep, and the dreamer will act out their dream.*

---

* In some cases – fortunately very rare – this connection with the body during sleep can be very severe. A dramatic example was a Welshman, Brian Thomas, a Good Samaritan and devout Christian, who, in the middle of a nightmare in which he believed he was fighting against a burglar, strangled his wife to death. When he woke up, devastated and confused, he called the police to tell them that he had murdered his life partner of forty years.

What does the brain do as we dream? The first thing we should know is that the brain does not turn off while we sleep. Actually, the brain never stops; if it shuts off, our life ends. When we are sleeping, the brain is carrying out sustained activity, both throughout the REM (rapid eye movement) phase during which we dream, and throughout slow wave sleep, which is deeper and typically without dreams.

The myth that the brain shuts off at night is tied to the idea that sleeping is a waste of time. We recognize the merits of our own and others' lives through the achievements we make while awake – the jobs, friends, relationships – but there is no merit in being a good dreamer.*

Sleep is a reparative state, during which a cleaning programme is carried out, eliminating biological waste and residue from the cerebral metabolism. In effect, the brain, over the course of the night, takes out the rubbish. This relatively recent biological discovery is in line with the common, intuitive idea that sleep is the functional flipside of our waking lives and without it, besides being tired, we get sick.

Beyond this restorative role, key facets of the cognitive apparatus are set in motion while we are sleeping. For example, during one of the first phases of sleep – slow wave – memory is consolidated. So after a few hours of sleep or even a quick nap, we remember better what we've learned over the course of the day. And that is not only due to the rest we've had. In fact, it is largely due to an active process that goes on while we are sleeping. By taking a closer look through experiments on the cellular and molecular level, we now know that during this phase of sleep specific connections between neurons in the hippocampus and the cerebral cortex that store and stabilize memory are reinforced. These changes begin during diurnal experience and are consolidated during sleep. This mechanism is so precise that, during sleep, it can recap exactly some neuronal patterns activated during

---

* Merits that could be duly compiled in a CS, a *curriculum somnii*.

the day. This is a contemporary physiological version of one of Freud's main ideas about sleep, the remains of the day. Those who are fans of naps could also argue that a long stretch of sleep at night is not necessary for this to be carried out. Short naps also function to consolidate memory.

During slow wave sleep, cerebral activity increases and decreases, forming repeating cycles over a period of little more than a second. In other words, the brain activity pulses oscillate in a clear, slow, defined rhythm. The more pronounced this oscillating wave of activity is, the more effective the memory consolidation. Can this oscillation be induced from outside the brain of the sleeper and thus improve their memory?

A person's rhythm of cerebral activity during sleep can be measured with an electroencephalogram. Then the neuronal activity of the sleeper can be increased, by making them hear sounds that are synchronized to the rhythm of their brain.

This experiment, carried out by the German neuroscientist Jan Born, began during the day with a list of new words that had to be remembered. Born discovered that people who later, during the night, listened to tones synchronized with the rhythm of their own cerebral activity would remember many more words the next day than those people who were not stimulated or were stimulated in a non-synchronized way.

This means that we can improve the memory of learning begun while awake by manipulating, in a relatively simple way, a cerebral mechanism that consolidates learning during sleep. However, the fantasy of putting on headphones at night and waking up speaking a new language that we've never studied continues to be just a fantasy.*

---

* *Désolé.*

The uroboros plot

Memory consolidation occurs during a phase known as slow wave, in which cerebral activity is monotonous and repetitive. But this is not the sole register of cerebral activity throughout sleep. In the REM phase, brain activity is much more complex and similar to activity while awake. In fact, during the REM period, the sleeper's subjective experience becomes conscious in the form of dreams.

When someone wakes up in the middle of the REM cycle, they almost always have a vivid memory of the content of their dreams. However, this does not happen when we wake up in other phases of sleep. From the point of view of our subjective experience, consciousness during sleep is similar to waking consciousness. In dreams we can fly, talk to people who are no longer alive, walk through a garden of half-buried train carriages, and even obey traffic laws. Dream images are vivid and intense. But, strangely, we lose the notion that we are the authors of the stories told in our dreams. We experience what we dream as if it were a true description of reality and not a figment of our imaginations.

The main difference between dream and waking consciousness is control. During sleep, as in schizophrenia, we do not detect our authorship of that virtual world. The bizarre nature of dreams is such that the brain does not recognize them for what they are: hallucinations.

While slow wave sleep is a state in which waking neuronal activity is repeated, during REM sleep more variable neuronal patterns are generated, with the ability to recombine pre-existing patterns of neural activity. Is this perhaps a metaphor for what happens on the cognitive plane? Is REM sleep a state conducive to creating new ideas and connecting elements of thought that were disconnected during the day? Are dreams a creative thought factory?

The history of human culture is filled with stories of revolutionary ideas originating from dreams. One of the most famous is that of August Kekulé, who discovered the structure of benzene, a ring of six carbon atoms. During a celebration of this great landmark in the history of chemistry, Kekulé revealed the secret behind his discovery. After failing miserably for years, the solution finally came to him as he dreamed of an uroboros, a serpent biting his own tail, making a ring shape. Something similar happened to Paul McCartney, who woke up in his bedroom on Wimpole Street with the melody to 'Yesterday' in his head. For days, McCartney searched in record stores and asked his friends for clues as to the origin of the melody, because he supposed that the dream had come from something he had listened to.

We can already anticipate the problem with these anecdotes: the conscious narrative is tinged with fiction. The same is true for memory, since we can recall with full conviction an episode that never happened. Even more extraordinary is that it is possible to implant a memory that a person then believes to be authentic. And invoking creativity during sleep can be a trick and a trap.

Perhaps with that hunch in mind, a chemist, John Wotiz, meticulously reconstructed the history of the discovery of the structure of benzene. And he found out that the French chemist Auguste Laurent, ten years *before Kekulé's dream*, had already explained that benzene was a ring of carbon atoms. Wotiz's thesis is that the invocation of the dream was part of Kekulé's strategy to hide his intellectual theft. What Paul McCartney honestly feared – that his dream was the expression of information gathered while awake – was deliberately manipulated by Kekulé.

But beyond the possible intrigue, can creative thought be shown to derive from dreams in an objective way that is not contaminated by the inevitable human distortions? That is what our dream hero, Jan Born, set out to discover.

The key to the experiment was finding an objective and precise way to measure creativity. In order to do that, Born set out a problem that could be solved in a slow but effective way or an original and simple one, by changing the perspective of the approach. The participants worked on that problem for a long while. Then, some slept and others just rested. Later, they all went back to their problem-solving. And the simple but conclusive result was that the creative solution was much more likely to appear after sleeping on it. Which is to say that part of the creative process is expressed while we sleep.

Jan Born's experiment shows us that sleep is an element in the creative process but not the only one. Despite a contemporary loss of prestige for drills and practice, the rote, ordered side of creativity is also important. Dreaming – like other forms of disordered thought – can help in the induction process of an original idea, but only after a firm base of great knowledge of the field in which we aspire to be creative has been established. We see this in the McCartney case; he had deep-rooted knowledge of the material and was later able to improvise in dreams. The same is true in Born's experiment. Night-time is the space of a creative process only after a day of arduous, methodical work that lays the foundations for creativity in dreams.*

That is how, in short, the thought factory works at full efficiency during the night shift. Sleep is a very rich, heterogeneous state of mental activity that allows us to understand how consciousness works. There is a first phase in which consciousness fades, not haphazardly but towards a place of great synchronization that activates a memory consolidation process. Then there is a second phase that is physiologically similar to the waking state but generates a more disordered pattern of cerebral activity. During this process an ingredient of creative thought is expressed, gestating new combinations and possibilities. All of this is accompanied by an oneiric narrative that can combine terror, eroticism and confusion. A full dream state. But are we really dreaming as we sleep? Or is it just one of our brain's many illusions?

---

* *A Hard Day's Night.*

Deciphering dreams

We have all had the experience of waking up thinking that we were only asleep for a few seconds, and actually hours have passed. Or the opposite, a few seconds of sleep sometimes seem to have lasted an eternity. As we sleep, time flows in an odd way. In fact, it's possible that the dream itself was *just* the illusion of a story constructed as we woke up.

Today we are able to resolve this mystery by observing traces of thought in the brain in real time. Just as we can investigate the thought processes of vegetative patients and babies, and subliminal processing based on cerebral activity, we can use similar tools to decipher our thoughts during dreams.

One way of decoding thought based on cerebral activity is dividing the visual cortex into a grid, as if each cell were a pixel in the sensor of a digital camera. Based on that we can reconstruct what is in the mind in the form of images or videos. Using this technique, Jack Gallant was able to re-create a strikingly clear film, by observing *only* the brain activity of the person watching the film.

This allowed a Japanese scientist, Yukiyasu Kamitani, to develop a kind of oneiric planetarium. His team reconstructed the plot of dreams based on the cerebral activity of the dreamers. Once they were awake, it was proven that the conjectures they'd based on their patterns of cerebral activity coincided with what the participants said they had dreamed.

They were narratives of this type: 'I dreamed I was in a bakery. I grabbed a baguette and went out on to the street, where there was someone taking a photo'; 'I saw a large bronze statue on a small hill. Below it were homes, streets, and trees.' Each one of these dream fragments was decoded on the basis of cerebral activity. In this demonstration the conceptual skeleton of dreams was deciphered, but not their visual qualities, their glimmering and their shadows. Reconstructing dreams in high definition and Technicolor is still being worked on in the experimental kitchen.

Daydreams

During sleep, the brain does not shut off but is actually in a highly active state, carrying out vital functions for a proper working of the cognitive apparatus. But it also happens that when we are working, driving, talking to someone or reading, our brain frequently unmoors from reality and creates its own thoughts. We often spend a large part of the day talking to ourselves. This is daydreaming, the expression of a state similar to dreams in form and content but while we are wide awake.

Daydreaming has a very clear neuronal correlate. While we are awake, the brain organizes itself into two functional networks that, to a certain extent, alternate. The first we are already familiar with: it includes the frontal cortex (which functions as the control tower), the parietal cortex (which establishes and links routines, controls space, the body and attention) and the thalamus (which functions as a traffic distribution centre). These nodes are the nucleus of a mode of active cerebral functioning that is focused and concentrated on a particular task.

When dreaming invades our waking state, this frontoparietal network deactivates and another group of brain structures takes control, near the plane that separates the two hemispheres. This network includes the medial temporal lobe, a structure linked to memory, which could be the fuel behind our daydreams. And also the posterior cingulate, which is highly connected to other regions of the brain and coordinates daydreaming just as the prefrontal cortex does when the focus is on the outer world. This system of brain regions is called the *default* mode network, a name which reflects how it was discovered.

When it became possible to explore human brain functioning in real time with functional magnetic resonance imaging (fMRI), the first studies compared cerebral activity while someone was doing something – a mental calculation, playing chess, remembering words, talking, expressing emotion – with another state in which *they were doing nothing*. In the mid-nineties, Marcus Raichle discovered that when a person is doing these tasks some regions activate while others deactivate. With one important distinction, the brain regions that activate vary depending on the task, while the ones that deactivate are always the same. Raichle understood that this reflected two important principles: (1) there is no such thing as a state in which our brain *does nothing*, and (2) the state in which thought wanders at its own volition is coordinated by a precise system which Raichle called the *default* network.

The structure of the brain's default network is almost diametrically opposed to the structure of the executive control network, reflecting certain antonymy between these two systems. The awake brain constantly alternates between a state with its focus on the outer world and another governed by daydreams.

Are daydreams just wasted time, some sort of cerebral distraction? Or maybe, like nocturnal dreams, they have a good reason for existing in the framework of our way of thinking, discovering and remembering.*

The way our thoughts sometimes drift as we read is fertile territory for the study of daydreaming. We have all had the experience of suddenly realizing we don't have the faintest idea of what we've been reading over the last three pages. We were occupied with a parallel story that pushed the contents of our reading to the margins of our consciousness.

A careful recording of eye movements shows that during daydreams we continue to scan word by word as we read, and to slow down on the longer words. But at the same time, during that daydreaming, activity in the prefrontal cortex lessens and the default

---

* Luis Buñuel clearly took a side in this debate: 'Dreaming while awake is as unpredictable, important and powerful as doing so while asleep.'

system activates, which makes the information from the text we're reading fail to access the privileged gardens of consciousness. Which is why we go back with the sensation that we have to reread the entire lost fragment again, as if it were the first reading. But that is not the case. This new reading builds on the previous one, done amidst dreams.

What happens is that as we daydream we read with a different focus, with a wide-angle lens that allows us to ignore small details and observe the text from afar. We focus on the forest and not the tree. Which is why when we daydream while reading and then go back to the same text, we understand it better than when merely skimming the text with full concentration. In other words, daydreaming is not that wasted time Marcel Proust so yearned after.

However, there are reasons to believe that daydreaming has a cost (that has nothing to do with the time it consumes). Dreams can easily turn into nightmares, hallucinations lead to bad trips and imaginary friends into monsters, bogeymen, witches and ghosts. Almost all of the situations in which the mind wanders and unhitches from reality can easily degenerate into states of suffering. This is an observation for which I do not have – and for which I do not yet believe there is – a good explanation. I can only share my hypothesis: the executive system, which controls the natural and spontaneous flow of thought, develops – in each of our personal cultural and evolutionary histories – to avoid this flow degenerating into suffering.

An American psychologist, Dan Gilbert, gave this idea physical substance with a cell phone app that every once in a while asks users: 'What are you doing?'; 'What are you thinking about?'; 'How are you feeling?' The answers, gathered from people throughout the world, comprise a sort of chronology and demographics of happiness. In general, the states of greatest happiness correspond to having sex, talking with friends, sports, and playing and listening to music, in that order. Those of least happiness are work, being at home at the computer or on public transportation in the city.

Obviously, these are averages and do not imply that working

makes everyone unhappy. And, naturally, these results depend on social and cultural idiosyncrasies. But the most interesting part of this experiment is how happiness changes according to what we are thinking about. During a daydream, almost all of us feel worse than when our brain isn't wandering freely. This doesn't mean that we shouldn't have daydreams but rather just that we should understand that they entail – like so many other trips – a complicated mix of discoveries and emotional ups and downs.

Lucid dreaming

Nocturnal dreams also often travel through painful and uncomfortable places. Unlike our imagination, dreams go *where they want to*, without our control. The other big difference between dreams and imagination is their visual intensity. Awake, we are scarcely able to reconstruct the tattered ruins of a vivid, intensely colourful dream.

So, dreams and imagination are distinguished by their degree of vividness and control. Dreams have no control but are vivid. Imagination, on the other hand, is controllable but much less vibrant. Lucid dreaming is a combination of both, it has the vividness and realism of dreams and the control of the imagination; in other words, it is a state in which we are both the director and scriptwriter of our dream. Given the chance to choose, most lucid dreamers want to fly, perhaps expressing an ancestral frustration of our species.

During lucid dreaming: the dreamers understand that they are dreaming; control what they dream; and can disassociate the object and the subject of the dream, as if they were watching themselves in the third person. And lucid dreaming also has its own cerebral signature. Brain activity during REM sleep has less high-frequency activity in the frontal cortex as compared to wakefulness. And it is precisely that high-frequency activity that is imperative in controlling lucid dreaming. In fact, the more lucid the dream, the greater the high-frequency activity in the prefrontal cortex. We could even flip that around. If the brains of normal dreamers are stimulated in high frequency, their dreams will become lucid. The dreamers will disassociate from their dreams, begin to control them deliberately, and will understand that they are just dreams.

A future in which we control our dreams is not far off. It won't even require that much technological fanfare. We have known for some time that the ability to have lucid dreams can be trained and that, with work, almost anyone can experience them. A way of approaching them is through nightmares, which we feel a natural desire to control. The capacity that many people have to direct the course of their nightmares – including the executive control to wake up – is a prelude to lucid dreaming. And vice versa: training lucid dreaming is a way of improving the quality of dreams. So another of its distinctive traits is a higher density of positive emotions.

As part of the training process, lucid dreamers use a waking world as an anchor, allowing them to know that they are in a dream, and that *on the other side* is the reality of wakefulness. This functions as an orienting reference point to understand where they are. Like Theseus, Hansel or Tom Thumb, or like Leonardo di Caprio in *Inception*, lucid dreamers leave trails through their wakefulness that serve to guide them back when the path through their dreams becomes too winding.

Lucid dreaming is a fascinating mental state because it combines the best of both worlds, the visual and creative intensity of dreams with the control of wakefulness. And it is also a gold mine for science. Gerald Edelman, a Nobel laureate, divides* consciousness into two states. There is a primary one, which constitutes a vivid story of the present, with very restricted access to the past and the future. It is the *Truman Show*'s consciousness, that of a passive spectator who sees the plot of his reality live and directly. This is, according to

---

* The timetable of a book is strange. The reader's present is the author's past. Gerald Edelman died in May 2014, after this page was written and before it was read. I chose to maintain the present tense, from the perspective of when Edelman was still expressing his ideas, which remained clear and thought-provoking to the end of his days.

Edelman, the consciousness of many animals and also that of REM sleep. A consciousness without a pilot. A second form of consciousness, richer and perhaps more personal to humans, introduces the necessary ingredients for the pilot to function as such; it is abstract and creates a representation of one's self, of one's being. Perhaps lucid dreaming is an ideal model for studying the transition between these primary and secondary states of consciousness. We are now in the first stages of sketching out this fascinating world that has only recently appeared in the history of science.

Voyages of consciousness

Another age-old path to social and personal exploration of our consciousness is the ingestion of medicines, plants, herbs, coffee, chocolate, tea, alcohol, cocaine, opium, marijuana . . . Substances that can be stimulating, psychoactive, hallucinogenic, soporific, hypnotic. Psychopharmacological exploration that seeks to link the effect of plants, their compounds, derivatives and synthetic versions with specific mental states has been an exercise common to every culture. Here we will explore the universe of the science of two types of drugs that alter the content and flow of consciousness: cannabis and hallucinogenic drugs.

The factory of beatitude

Cannabis is a plant native to South Asia that has been used to make clothes, sails, riggings and paper since at least 5,000 years ago. The use of cannabis as a drug* is also a practice that dates back millennia:

---

* While a single word, *dream*, can refer to many different things – an oneiric representation, an aspiration, a fantasy – *marijuana* is a case of the opposite process, born from the taboo and shame of naming something. So a single meaning is expressed by a multitude of words: pot, weed, grass, ganga, dope, herb, chronic, reefer, skunk, Mary Jane.

this may explain why a shaman from the Xinjiang region of China was mummified alongside a basket containing cannabis leaves and seeds. There are also records of the use of cannabis in Ancient Egyptian mummies and icons.

In the 1970s laws prohibiting the recreational and medicinal use of cannabis burgeoned and, approximately forty years later, the wave began to recede. In effect, the legality of a drug changes abruptly from one place and time to the next, and in general this decision ignores the mechanisms and details of its biological functioning. In order to be able to make an informed decision, whether public or private, one needs to know how different drugs affect the brain and the mind. This is currently particularly relevant in the case of marijuana, when its legalization is being extensively debated.

In the seventies, the three most widely used illegal recreational drugs were marijuana, opium – as morphine and heroin – and cocaine. The psychoactive compounds in opium and cocaine, as well as their primary mechanisms of action, had already been identified. Practically nothing was known about marijuana. After earning his doctorate at the Weizmann Institute and doing a post-doc at Rockefeller University, a young Bulgarian chemist, Raphael Mechoulam, returned to Israel to remedy this ignorance. Establishing the bridge between the chemical molecules in cannabis and its action on the body and mind was already a significant statement:

I believe that the separation of scientific disciplines is just an admission of our limited ability to learn and understand several scientific areas. In nature, the border does not exist.

This impressive declaration of intent defined Mechoulam's research style. This book is, to a certain extent, an heir to that legacy.

His road was not – and is not – an easy one, largely because of the illegality of the substance he wanted to study. In order to work he had to employ tricks unimaginable to most researchers. First of all, he had to get the cannabis. Taking advantage of his military experience, Mechoulam convinced the Israeli police to let him acquire five kilos of Lebanese hashish to begin his long project. His task next was to chemically isolate the almost one hundred compounds that make up cannabis and then give them to monkeys to identify which were responsible for the psychoactive effects. Since it is not easy to tell when a monkey is stoned, he used the sedative effect as a marker to determine the potential of each compound. And thus, in 1964, he managed to identify Δ1-tetrahydrocannabinol (Δ1-THC, now known as Δ9-THC) as the primary compound responsible for cannabis's psychoactive effects. Other compounds that are much more frequently found in marijuana, such as cannabidiol, have no psychoactive effect. However, they have physiological effects as anti-inflammatories and vasodilators and are, in fact, the primary focus of its medicinal uses.

Discovering the active compound of a plant is just the first step to being able to investigate its mechanism of action. What happens in the brain that sets off the explosion of appetite and laughter and changes in perception? Mechoulam's second big discovery was identifying a receptor in the brain that specifically reacts to Δ9-THC. A receptor is a molecular sensor on the surface of a neuron. The active compound of the drug is like a key for which the receptor is the lock. Of all the locks in the brain, Δ9-THC only opens a few, which are called the cannabinoid receptors. We now know of two types: the CB1, distributed in neurons in a wide range of cerebral regions, and CB2, which regulates the immune system.*

When a molecule fits into a receptor on the surface of a neuron it can produce different changes in that neuron: activate it, deactivate it, make it more sensitive or change the way it communicates with its neighbouring neurons. This happens simultaneously in the millions of neurons that have this type of receptor. On the other hand, this

---

* We know that there are more receptors, although they have yet to be found, because when the CB1 and CB2 are blocked (when the locks are covered up), cannabis continues to produce physiological and cognitive effects.

molecule does nothing to the neurons that do not have a receptor that reacts to Δ9-THC.

Molecules and their receptors do not fit together perfectly. The key sometimes fails to open the lock. The better the fit, the more effective and powerful the drug response. By studying the chemical structure of cannabis, Mechoulam was able to synthesize a compound a hundred times more effective than Δ9-THC. Five grams of that compound produce an effect equivalent to some twenty-five pounds of marijuana.

Why do the neurons of the human brain have a specific receptor for a plant that grows in South Asia? It is strange that the human brain has a mechanism to detect a drug that for centuries only really grew in very specific parts of the planet. Does this system have no use for those who do not consume cannabis? Were these receptors, which are so prominent in the brain, just sitting there unused until marijuana became popular?

The answer is no. The cannabinoid system is a key regulatory piece of the brain for us all, regardless of whether we smoke weed or not. The solution to this enigma is that the body manufactures its own version of cannabis.

In 1992 – almost thirty years after the discovery of THC (science cooks on a slow flame) – Mechoulam (older but no less persistent) made his third big finding: an endogenous compound that the body produces naturally and that has the same effect as cannabis. He called this compound anandamide, because it is an amide (chemical compound) that produces *ananda*, which in Sanskrit refers to beatitude.

This means that each and every one of us, in the opaque and intimate silence of our physiology, creates cannabis. The activation of the cannabinoid receptors from marijuana consumption is much greater than what is naturally produced by anandamide. The same is true of almost all drugs. The endorphins (endogenous opioids) that we normally produce in our bodies – for example, when running – activate our opioid receptors much, much less than morphine or heroin.

This distinction is key. In many instances, the fundamental difference between two compounds is not found in their mechanism of action but rather in their dosage. For example, Ritalin and cocaine act by exactly the same mechanism. The first is legal and is used to treat attention deficit disorder in children. Leaving aside the discussion of its possible medical abuse, it is clear that Ritalin does not have nearly the addiction potential of cocaine. The reason behind this fundamental difference is entirely due to its concentration.*

The cannabic frontier

The CB1 cannabis receptor is found throughout the entire brain. This distinguishes it from dopamine receptors (for cocaine) that are only in specific parts of the brain. This means that many neurons in different cerebral regions change their function following marijuana consumption. Today we have detailed information on some aspects of cannabis's biochemistry. For example, some neurons known as POMC, which are found in the hypothalamus, produce a hormone that regulates satiety and suppresses appetite. But when the CB1 receptor is active it opens up a structural change in the neuron that makes it manufacture a different hormone with the opposite effect, stimulating the appetite. A close-up biochemical look at the brain's hormone factory explains this effect known to all pot smokers: the munchies, a voracious hunger that doesn't wane however much they eat.

While the relationship between marijuana and appetite is known in exquisite detail, the bridge between the biochemistry, physiology and psychology of the drug's cognitive effects remains a mystery. Those who smoke or ingest marijuana have the sensation that their consciousness changes. How can science investigate such a subjective

---

* It is Paracelsus's formula, valid since the fifteenth century: the only difference between a poison and a remedy is the dose.

aspect of perception? I am not referring to how much we remember or how fast we can add after smoking, but rather to a much more introspective question. The reorganization of thought after consuming cannabis is a mystery that science has barely scratched the surface of.

The lack of scientific information on the cognitive effects of cannabis is due, first of all, to marijuana's illicit status. Mechoulam's road was an exception in this abyss of ignorance. And finding consensus in the relatively scant scientific literature is not an easy task either. A search rapidly reveals contradictory results: that marijuana affects memory and that it doesn't. That it radically changes one's ability to concentrate and that it doesn't alter it in the slightest.

We are not accustomed to such dissent in scientific literature, but actually it is not limited to this field. To give a non-pharmacological analogy, is it good or bad for a kid to spend hours playing on the computer? Parents who want to find information and duly regulate access to screens will find a hot mess. One work will recognize the benefits of games on cognitive development, attention and memory; another will warn of its detrimental effects to social development, and so on.

This dissonance has several explanations. The first is that there is not just *one* marijuana – there are many different kinds. The concentration and the ingredients vary (more or less THC), but also the ways of consuming it, the quantities and the user's metabolism. To give a much more straightforward example, it's like trying to resolve across the board whether eating sweets is good or bad. It depends on how much sugar they have in them, what kinds of sugars they contain, and who is eating them, i.e. whether they are obese or diabetic or very skinny and hypoglycemic.

The fact that there are studies with such varied conclusions suggests that the possible risks of marijuana are not universal. On the other hand, if we take all the scientific literature as a whole, we see that there is a consistent finding that marijuana poses the risk of inducing psychosis in teenagers or people with prior psychiatric pathologies, both when smoking it and for some time afterwards. In fact, an effect common to most drugs, not just marijuana, is that the age when consumption begins enormously affects its addictive potential. The younger the user is at initial consumption, the more likelihood there is of that substance becoming addictive.

Towards a positive pharmacology

There is a fine line between relieving pain and seeking pleasure, even if later society builds an abrupt wall atop this fine line. It is usually considered acceptable to pump someone in pain full of drugs yet deny the slightest use to someone who is *fine* but wants to feel a bit better. This asymmetry also occurs in science, which focuses on the detrimental effects of marijuana and largely ignores its possible positive effects.

Practically all the scientific research on marijuana has to do with determining whether it distances us from some presumed normality. On the other hand, it is hard to find works that investigate whether the line demarcating normality could be moved to a better place. Something similar was seen in psychology; little more than thirty years ago, it was only concerned with improving the condition of those who were depressed, anxious or frightened. Martin Seligman and other researchers changed the focus by founding positive psychology, which deals with research on how to make that *normal* better.

Science would be much more honest if it could also create a positive pharmacology. This path was explored in literature with Aldous Huxley as a standard-bearer, in *The Doors of Perception*, but was almost ignored by scientific inquiry. A possible path of investigation could begin by not thinking of marijuana only in terms of whether it is harmful but whether it can be used to live better. This, obviously, does not mean that marijuana is good. The challenge is in discovering to what extent it can improve everyday life; for example, making us laugh more, socialize and enjoy more, or have better sex. Basically the idea would be to weigh that against the real risks – which exist and in some cases are severe – in order to be able to make better decisions, in both the private realm and the public, political one.

The consciousness of Mr X

Carl Sagan, the author of *Cosmos* and one of the most extraordinary popular science writers ever, smoked marijuana for the first time when he was already an acclaimed scientist.* As usually happens, his first experience was a fiasco, and Sagan, a battle-hardened sceptic, came up with all sorts of hypotheses about the drug's placebo effects. However, according to Mr X – his cannabic alter ego – after a few more attempts, the drug took effect:

> I viewed a candle flame and discovered in the heart of the flame, standing with magnificent indifference, the black-hatted and -cloaked Spanish gentleman [. . .]. Looking at fires when high, by the way, especially through one of those prism kaleidoscopes which image their surroundings, is an extraordinarily moving and beautiful experience.

In Mr X's account, he didn't confuse this manipulation of his perception with reality, exactly like in a lucid dream:

> I want to explain that at no time did I think these things 'really' were out there. I knew there was no Volkswagen on the ceiling and there was no Sandeman salamander man in the flame. I don't feel any contradiction in these experiences. There's a part

---

* The relationship between drugs and professional success can also take the opposite route. A text that some believe to be apocryphal tells the story of Adrián Calandriaro, who after composing two highly imaginative records sought to resolve a long period of musical drought and locked himself up with a notebook, a pen and 2,000 hits of lysergic acid. Calandriaro stayed tripping from 14 May 1992 to mid-April 1998. In that period he studied odontology, set up a consulting firm, married, and had two kids, a dog named Augusto and $2,000,000 in a bank account in Uruguay. He's happy, but misses music a bit (*Peter Capusotto*, the book).

of me making, creating the perceptions which in everyday life would be bizarre; there's another part of me which is a kind of observer. About half of the pleasure comes from the observer-part appreciating the work of the creator-part.

His change in perception from cannabis was not exclusive to the realm of images. In fact, the most substantial modification was probably to his auditory perception.

For the first time I have been able to hear the separate parts of a three-part harmony and the richness of the counterpoint. I have since discovered that professional musicians can quite easily keep many separate parts going simultaneously in their heads, but this was the first time for me.

Mr X was also convinced that the ideas which seemed brilliant under the effects of cannabis were truly brilliant. Sagan tells of how, in fact, part of the more laborious and methodical work he did in his life was ordering those ideas, recording them on tape or in writing – at the cost of losing many other ideas – and that the next day, when the effects of the marijuana had passed, the ideas had not only not lost their appeal but cemented a large part of his career.

A neuroscientist colleague and friend of mine – let's call him Mr Y – carried out an informal and personal project inspired by Carl Sagan's account. The experiment consisted in observing, while high on marijuana, an image that vanished very quickly. Then he had to indicate what was in different fragments of the image and the vividness with which he remembered it.

Without smoking he was able to recall a small fraction of the image, due to the narrowness of consciousness. But high, Mr Y believed he remembered it all with great clarity and had the sensation that he was discovering something extraordinary and

151

singular. He felt like he was in Huxley's head, opening the doors of perception.

When the experiment was over, he anxiously but carefully analysed the data to discover that actually he had seen exactly the same thing after smoking as he had before. Exactly the same thing. It was the subjective landscape, how he felt the details of the image, that changed. Like Sagan, he felt a certain brilliance in his perception during the cannabic state, the same feeling that makes us overestimate the funniness of a joke or the originality of an idea.

This experiment and Sagan's coincide in the subjective richness of the cannabic state but differ in whether it is genuine or a figment of the imagination. Settling between these two alternatives turns out to be impossible because, unlike the rest of the experiments described in this book, these ones lack the necessary scientific rigour to be able to draw firm conclusions. This is mostly because of how hard it is to experiment with cannabis in a rigorous way.

One of the most informative studies on how the brain re-organizes as a consequence of extended cannabis use was published in *Brain*, one of neuroscience's most prestigious magazines. The attention and concentration abilities of frequent chronic smokers – on average, they'd smoked more than 2,000 joints – were studied and compared to people who had never smoked marijuana. In this case, attention was measured by see-ing how many points they were able to follow at the same time, without mixing them in their minds and without losing track of which was which. In other words, a mental juggling exer-cise. The result of the study was that smokers and non-smokers have a very similar attention capacity and resolve the problem with more or less the same skill. Therefore, the first conclusion was that cannabis users, on average, do not lose or gain in their ability to pay attention and concentrate.

The most interesting finding was that, despite this similarity in their performance, the cerebral activity of each of the two groups is very different. Cannabis users activate less their frontal cortex – which regulates mental effort – and their parietal, while activating more their occipital cortex – the territory of the visual system that functions like the brain's blackboard. The change in cerebral activity between those who smoke and those who don't – more occipital activity, less frontal – is similar to what's observed when comparing the cerebral activity of great chess masters and novices while they play. The chess masters activate the occipital cortex more and the frontal less, as if they were seeing the moves instead of calculating them.

This result has two possible interpretations. One is that those who smoke marijuana activate the frontal cortex less because they don't need to use so much effort to resolve the problem, like the chess master playing someone easy to beat. The other possibility is that their attention system is compromised and they use their visual cortex more to remedy and compensate for this lack. The difference is subtle but opportune. Studying it carefully could allow us to separate the risks from the benefits and understand how they balance each other out in a mental state that is not necessarily better or worse than the *normal*, just different.

The lysergic repertoire

Ayahuasca is the most celebrated potion in the Amazonic world. It is served as a tea prepared with a mix of two plants, the *Psychotria viridis* bush and the *Banisteriopsis caapi* vine. Actually there are different formulas, but they all include these two plants with neuropharmacologically complementary roles. The bush is rich in N,N-Dimethyltryptamine, better known as DMT. The vine has a monoamine oxidase inhibitor (MAOI), one of the most commonly used antidepressants.

There is a synergy between these two drugs that make up ayahuasca. The DMT modifies the neurotransmitter balance. In normal situations, monoamine oxidase, like the brain's chemical police, would resolve this imbalance. But here it interacts with the MAOI of the vine, which inhibits the brain's ability to regulate its neurotransmitter balance. So, in the dosage used in ayahuasca, the psychedelic effect of the DMT is low, but its combination with the vine makes it stronger. Ayahuasca radically changes perception and induces severe transformations in the pleasure and motivation systems. Of course, it also alters the flow, organization and anchor of consciousness.

Of all the perceptive changes ayahuasca produces, the most extraordinary are very vivid hallucinations called *mirações* (visions). They are highly visual constructions built by the imagination. Under the effects of ayahuasca, the imagination has the same resolution as vision. How does this materialize in the brain?

Draulio Araujo, a Brazilian physicist accustomed to trekking through jungles and wetlands, did a unique experiment that syncretized ancestral traditions of the Amazon region and the most sophisticated technological development. Draulio brought shamans, experts in using the potion, to the modern, aseptic rooms of the hospitals in Riberão Preto so they could take the drug and then enter into the resonator to give their visions full rein.

There, in the intimacy of the resonator, the expert shamans hallucinated and then reported on the intensity and vibrancy of their hallucinations. They later repeated the experiment without the effects of the drug, when the imagination is expressed in a much more subdued way.

When we see an image, information travels from the eyes to the thalamus, then to the visual cortex and from there to the formation of memories and the frontal cortex. With ayahuasca, the visual cortex is fed not by the eyes but by the inner world. It is this reversal of the flow of information that underlies the hallucinations. During a psychedelic hallucination, the circuit begins in the prefrontal cortex and from there feeds off the memory to flow backwards to the visual cortex. The chemical transformation of the brain results – through mechanisms we have yet to discover – in a projection of the memory on the visual cortex, as if reconstructing the sensory experience that gave rise to those memories. In effect, on ayahuasca, the visual cortex activates with practically the same intensity when seeing something as when imagining it, and that gives the imagination more realism. Without the drug, the visual cortex activates much more when seeing than when imagining.

Ayahuasca also activates the Brodmann Area 10, which forms a bridge between the external world – that of perception – and the inner world – that of the imagination. This explains an idiosyncratic aspect of the effects it produces. Commonly, those who take ayahuasca feel that their bodies are transforming; they literally feel that they are outside their own bodies. The border between the outer world and the inner world becomes more fuzzy and faint.

Hoffman's dream

In 1956, Roger Heim, director of the National Museum of Natural History in Paris, made an expedition to Huautla de Jiménez, Mexico, with Robert Wasson to identify and collect mushrooms used in the healing and religious rites of the Mazatecs. When he returned to Paris, Heim contacted the Swiss chemist Albert Hoffman to help him identify the chemistry of the sacred mushrooms. Hoffman was the ideal candidate for this task. Ten years earlier, after accidentally ingesting 250 micrograms of a lysergic acid he had just synthesized in his lab, he had the first acid trip in human history while riding home on his bike.

While Hoffman was figuring out that the magic molecule in the mushrooms was psilocybin, Wasson published an article in *Life* magazine entitled 'Seeking the Magic Mushroom', in which he gave an account of his trips to the Mexican desert with Heim. The article was a hit, and psilocybin ceased to just be a Mazatec object of worship and became a massive icon of Western culture in the sixties.

Lysergic culture would have a great impact on the Beat Generation of intellectuals such as Allen Ginsberg, William S. Burroughs and Jack Kerouac, who founded a movement that sought to radically change aspects of culture and human thought. Timothy Leary, with his Harvard Psilocybin Project, accompanied the lysergic generation by heading a scientific exploration of the transformative effects of psilocybin.

The three men who were at the origins of the science of psilocybin had central roles in science, economics, politics and culture. Wasson was vice-president of JP Morgan; Heim was decorated with the title of High Official of the Legion of Honour among other important French titles, and Hoffman was a senior executive at Sandoz, one of the top pharmaceutical companies, and a member of the Nobel Prize committee. However, in a certain sense, at least from the perspective of their extremely ambitious founding objective, the lysergic generation was a failure.

The spike of enthusiasm of a decade of research was followed by almost half a century of lethargy, in which psilocybin almost completely disappeared from scientific exploration, or at least was very marginalized. In the last few decades, curiosities of the mind were considered acceptable if they stemmed from dreams or peculiar brains, but the pharmacological exploration of the fauna and diversity of the mind came almost to a complete halt. However, this is changing. And this is to a great extent the consequence of a heated debate about drugs, politics, psychiatry and science that has taken place in Britain over the last ten years.

It begun when David Nutt, then in Bristol and now a Neuropsychopharmacology professor at Imperial College, London, was appointed in 2008 chairman of the Advisory Council on the Misuse of Drugs. From this position of great prestige and responsibility, Nutt begun a quite ferocious discussion with government officers about the criteria used to assess harm and dictate policies of drug use and abuse.

He championed the views that (1) drug legislation has to be set based on the best evidence we have available to determine the harm they cause, and (2) avoiding all-or-none arguments of harm in drugs and shifting instead to a more quantitative argumentation of the extent, scale and type of harm. To this end he conceived a taxonomy of several parameters which measure different dimensions of the negative consequences of a drug: physical harm, dependence, social harm . . . On the basis of this classification, he and his colleagues came to the conclusion that some legal drugs, like alcohol or tobacco, are more harmful than some illegal drugs like LSD, ecstasy or cannabis. The specific case of cannabis led to a big public and political confrontation when the government, ignoring these recommendations, changed its status from Class C to Class B (which implies that it is classified as a drug with greater risks that needs to be controlled more severely). On the public media and in the scientific and medical journals, Nutt argued that this decision was only politically motivated and was against scientific and evidence-based argumentations.

In a very celebrated and controversial article Nutt made his point of comparing the real danger of drugs to other domains of life in which we accept taking some degree of informed risk to seek pleasure. Nutt quantified the risk (in terms of physical harm, its addiction, how it distracts from work and can put a family in economic danger . . .) of *equasy*. After showing that the risks of this new drug were comparable to those of ecstasy, Nutt revealed the secret of the overlooked addiction of *equasy*: it was simply horse riding.

After all these heated discussions, in a very controversial decision, Nutt was dismissed from his position in the council by the Home Secretary. Since then Nutt has continued his effort to establish an evidence-based and rational discussion on drug harm and use. But he also came back with full force to the laboratory where some years later he met my friend and colleague Robin Carhart-Harris. And together they took up the baton that had been relinquished a few decades ago by Wasson, Heim and Hoffman and began a full new programme to understand how cerebral activity is organized during a psilocybin trip.

Now, in David Nutt's laboratory, there are all sorts of experiments on how cerebral activity is organized during a psilocybin trip. The Mazatec and Amazonic ritual traditions differ in which plants are used (mushrooms instead of vines and bushes), in their drug content (psilocybin instead of DMT and MAOI), in the type of psychological transformation, and also in the cerebral reorganization after the drug is ingested.

Psilocybin changes the way cerebral activity is organized in space and time. The brain spontaneously forms a sequence of states. In each one a certain group of neurons activate and later deactivate to give way to a new state, like moving clouds which make a figure and then dissipate to give rise to new shapes. In this metaphor, each group of clouds in a defined shape corresponds to a cerebral state. The succession of cerebral states represents the flow of consciousness. Under the effects of psilocybin, the brain travels through a greater number of states, as if the wind were making the clouds mutate more quickly in a much more varied repertoire of shapes.

The number of states is also a footprint of consciousness. During unconsciousness – deep sleep or anaesthesia, for example – the brain collapses into a very simple mode, with few states. When consciousness is switched on, the number of states increases, and with the induction of psilocybin that number grows even more. This could explain, from the brain, why many people who consume LSD and psychedelic mushrooms perceive a form of expanded consciousness.

In a lysergic state, many also mention something known as *trailing*, in which reality is perceived as a series of fixed images that drag a trail along behind them. So, with psychedelic mushrooms, the doors of perception, in addition to opening, become fragmented. The curtain is lifted, showing that the reality we perceive as a continuum is a mere succession of images – that property which Freud conjectured Ω neurons have in order to be able to both persist and change, the way consciousness does.

During normal perception, reality appears to be continuous, not discrete. But this is an illusion. As mentioned earlier, the discrete character of *normal* perception is subtly revealed in a car race. There we can frequently see a curious illusion in which the car wheels seem to turn backwards. The explanation of this phenomenon is well known in the world of film and television and has to do with the frequency of photographic stills that are being used to tell the story of reality. Imagine that the wheel takes seventeen milliseconds to go around once and that the camera captures a frame every sixteen milliseconds. Between one frame and the next the wheel has almost made a complete turn, so in each successive still the wheel appears to have gone slightly backwards. What's extraordinary is that this illusion is not an effect of the television screen but rather it is in our brain. This indicates that, as in film, we generate discrete frames that we later interpolate with an illusion of continuity. Perception is always fragmented, but only under the effects of a drug like psilocybin does this fragmentation become evident. As if we were seeing reality as it is behind the curtain, in the background of the matrix.

The past and the future of consciousness

Today, using the tools that allow us to infer thoughts from brain activity, it is possible to delve into dreams, into the mind of newborns and into the imagination of vegetative patients. But this technology is useless for investigating one of the most mysterious aspects of human thought: our predecessors' consciousness. We know for certain that their brains were almost identical to ours. But in our prehistory there were no books, radio, television or cities. Life was much shorter, and focused on hunting and the vital matters of the present. Was consciousness different from ours in contemporary society? To put it another way, does consciousness – the way we perceive it today – emerge spontaneously in the brain's development or is it forged in a particular cultural niche?

We may all have different opinions and intuitions about this; actually it is a long-standing philosophical debate. When I first reflected about this question, I believed that it was not even amenable to science. But then it became quite obvious that in the same way that we can reconstruct what the Ancient Greeks' cities looked like, on the basis of just a few bricks, the writings of a culture are archaeological records, fossils of human thought.

And in fact, doing some form of psychological analysis on a number of the most ancient books in human culture, Julian Jaynes sought to answer the question, coming up with one of the most polemical and debated theories in cognitive neuroscience: that only 3,000 years ago the world was a garden of schizophrenics. That consciousness, as we perceive it today, where we feel we are the pilots of our own existence, emerged with culture only very recently in the history of humanity.

The proliferation of the first books, in a period between 800 and 200 BC, coincided with radical transformations in three great civilizations in the world, Chinese, Indian and Western. It was during this period that the religions and philosophies that are the pillars of modern culture were produced. Studying two foundational texts of Western civilization, the Bible and the Homeric sagas, Julian Jaynes argued that consciousness was also transformed during this period.

He made this claim based on the fact that the first humans described in these books behaved – in different traditions, in different places of the world – as if they were hearing and obeying voices . . . which they perceived as if they were coming from gods or muses. What today we would call hallucinations.

And then, as time went on, they progressively began to understand that they were the creators and owners of these inner voices. And, with this, they acquired introspection: the ability to think about their own thoughts.

The Canadian philosopher Marshall McLuhan argued that this change was a consequence of the appearance of written texts, because it allows thought to be consolidated on paper instead of being entrusted to the more volatile memory. Those who now reflect so much on how

the Internet, tablets, mobile phones and the unceasing flow of information can change the way we think and feel should remember that
the information age is not the first material revolution to radically
change the way we express ourselves, communicate and, almost certainly, think.

For Jaynes, consciousness, prior to Homer, lived in the present and
didn't recognize that each of us is the author of our own voices. That
is what we call primary consciousness, which today is characteristic of
schizophrenia or dreams (except for lucid ones). With the proliferation of texts, consciousness transformed into what we now recognize.
We feel we are the authors, protagonists, and those responsible for our
mental creations, which in turn have the richness to interweave with
what we know of the past and what we predict or hope for the future.
And we are capable of introspection: we can think and reflect on our
thoughts.

When I first heard about Jaynes's theory, I thought that it was
quite spectacular, because of its capacity to put some order in the history of our own thoughts, and its wild conjecture that consciousness
may have been experienced in a completely different way at different
moments of our history. But it had an obvious problem. It was based
on just a few very specific examples and felt a bit like drawing constel-
lations in the sky.

With Guillermo Cecchi, my brother for many years in this adventure in science, and with two computer scientists, Carlos Diuk and
Diego Slezak, we decided to work out how this hypothesis could be
examined in a quantitative and objective manner. And the problem of
how to go about this is quite obvious: it is not like Plato woke up one
day and wrote: *Hello, I'm Plato, as of today I have a fully introspective
consciousness*. And this tells us what the essence of the problem is. We
need to find the emergence of a concept that is never said. The word
introspection is not mentioned one single time in the books we were
analysing.

Our way to solve this was to construct the space of words: a very
complex space in which all words are arranged so that the proximity

of two words is indicative of how closely related they are. We would want in this space the words 'dog' and 'cat' to be close together, but the words grapefruit and logarithm to be very far apart. And this has to be true for any two words in this space.

There are different ways to build this space. One is to ask the experts, just as we do with dictionaries. Another way is to follow the simple assumption that when two words are related, they tend to appear in the same sentence, paragraph or document more frequently than they would be expected to just by pure chance. And actually this simple method, with some computational tricks to deal with this huge and high-dimensional space, turns out to be very effective.

Once we have built this space, the question of the history of introspection, or of any other concept, which seemed abstract and somehow vague, becomes concrete, and amenable to quantitative science. All that needs to be done is to take a text, digitize it, project the stream of words in a trajectory on to this space and ask whether it spends significant time circling closely to the concept of introspection. The word *introspection* may never be said, but if words like self, guilt, reason, emotion become frequent then the text will be closer to introspection. This is how the algorithm can read in between the lines.

And with this we could analyse the history of introspection in the Ancient Greek tradition for which we have the best available written record. We took all books, ordered them by time, measured the proximity of each word to introspection, and calculated the averages, and then we were able to show that as time goes on there is a slow progression for the older Homeric texts: the *Iliad* and the *Odyssey*. And then, about 600 years before Christ, throughout the development of the Ancient Greek culture, the frequency begins to rise very rapidly to an almost fivefold increase and writings become closer and closer to introspection.

And the nice thing about using an objective procedure is that we can ask whether these results are also true in a different and independent tradition. So we repeated this analysis for the Judeo-Christian books, and we saw essentially the same pattern: a slow progression through the Old Testament with words getting closer and closer to introspection, and their use growing more rapidly throughout the course of the New Testament. Introspection peaks in the writings of Saint Augustine, about four centuries after Christ.*

This is very important, because Saint Augustine is widely recognized by scholars as one of the founders of introspection. (Actually some consider him to be one of the fathers of modern psychology.) So this algorithm, which has the virtue of being objective, quantitative and, of course, incredibly fast, can capture some of the most important conclusions of a long tradition of investigation.

One of the most relevant consequences of having converted this quest into objective science is that this idea can be translated and generalized to a whole range of different domains. And in the same way we used it to ask about the past of human consciousness, maybe the most challenging question we can ask is whether it can say something about the future of our own consciousness.

To put it more precisely, can the words we say today tell us something about where our minds will be a few months or even a few years from now? In the same way that many of us wear sensors (that detect our heart rate, our respiration, our genes), hoping that this information may help us prevent diseases, we can ask whether monitoring

---

* It is not possible to settle whether this change reflects the filter of written language, censorship, narrative trends and styles, effects of translations, or rewritings and new editions of the original books. There is also the issue of whether these books reflected popular thought or just that of the elites. And also whether these books are historical (and hence reflect real characters) or are simply fiction. There are many possible critiques which in my view are completely founded and inevitable in this form of research, in which thoughts are inferred from sparse and scattered traces left by our ancestors. The method we developed can prove what we call the soft Jaynes hypothesis: that as time goes on ancient books reflect more and more introspective content. It cannot go beyond this to provide direct evidence in favour of Jaynes's strong hypothesis, that this transition in text is a reflection of the way our ancestors thought. Resolving this dilemma requires ideas and tools that we have yet even to imagine.

and analysing the words we speak, we write, we tweet, may enable us to detect ahead of time when something might go wrong with our minds.

Guillermo Cecchi, in IBM Watson, put together a group of psychiatrists and computer scientists that spread from New York to Brazil and Argentina (often referred to humorously as the *Armada Brancaleone*) to take on this challenge.

We analysed the recorded speech of thirty-four young people who were at a high risk of developing schizophrenia. The question was whether properties of speech at day one could predict the onset of psychosis within a window of almost three years.

It turned out that there was just not enough information in semantics to predict the future organization of the mind. But this, in fact, was expected. One of the most distinctive features of schizophrenics is disorganized speech. Hence, the most important thing was not what they were saying, but how they were saying it. More precisely, it did not matter so much in what semantic neighbourhood the words were, but instead how far and fast they jumped in fluent speech from one neighbourhood to another. So we came up with a measurement that we termed semantic coherence, which grades the persistence of speech in one semantic topic.

And it turned out that in our group of thirty-four participants the algorithm based on semantic coherence could predict with close to perfect accuracy who would develop psychosis and who would not, something that could not be achieved by any other existing clinical measure. This is as yet a preliminary study of a relatively small group and needs to be replicated on a much larger cohort to calibrate its real efficacy, and the conditions in which it is most effective (how much speech, oral or written, structured interview or free . . .).

I was asked in 2016 to give a TED talk about this work. On preparing the talk I remember very vividly one day on which I saw a long series of tweets from Polo, one of my students in Buenos Aires who at the time was living in New York. There was something about these tweets. I could not tell exactly what, because they did not express anything explicitly. But I had a strong hunch, a strong intuition, that something was going wrong. So I took the phone and called Polo, and in fact he was not feeling well.

And this simple fact, that reading in between the lines I could sense through words his feelings, was a simple but effective way to help. In a way, I like to think that the most relevant aspect of this work is that it gets us closer to understanding how to convert this intuition, that we all have, that we all share, into an algorithm. And with this we may be seeing in the future a very different form of mental health, based on an automated, objective and quantitative analysis of the words we write, of the words we say.

The future of consciousness: is there a limit to mind-reading?

Today Freud would no longer be working in the dark. We have tools that allow us to access the thoughts – conscious or not – of vegetative patients and babies. And we can investigate the contents of a dreamer's dream. Will we soon be able to record our dreams and visualize them while awake, as in a film, in order to reproduce everything that, up until now, vanishes upon waking?

Reading others' minds by decoding mental states through their corresponding cerebral patterns is like tapping a phone line, cracking a code and entering into someone else's private world. This possibility opens perspectives and possibilities but also dangers and risks.* After all, if anything is private, it's our thoughts. Soon, perhaps, they no longer will be.

The resolution of today's tools is limited and *barely* allows us to recognize a few fragments of thought. In a not-so-distant future it is possible we could be able to write and read sensations directly from the biological substratum that produces them: the brain. And we will

---

* 'Never let anyone know what you're thinking' (Michael Corleone).

almost certainly be able to observe the contents of the mind deep in the most remote corners of the unconscious.

This path seems to have no end, as if it were only a question of improving the technology. Is that how it will pan out? Or is there a structural limit to our ability to observe our own and others' thoughts? In nature, as we know it, there are limits to our ability for observation. We cannot communicate faster than the speed of light, no matter what the technology. Nor can we, according to the laws of quantum mechanics, access all the information in a particle – not even its position and its speed – with absolute precision. Nor can we enter – or, more accurately, exit – a black hole. These are not temporary problems due to a lack of the proper technology. If our current knowledge of physics is correct, these limits are insurmountable despite any technological development. Will there be a similar limit to our ability to observe our own thoughts?

My friend and colleague the Swedish philosopher Kathinka Evers and I argue that there is a natural limit to our inspection of the human mind. The adventure can be extremely enriching – in some cases, liberating, as with vegetative patients – but it is likely that there is an intrinsic limit to our ability to investigate thought, one that goes beyond the limits of the technological precision of the tools with which we are examining it.

There are two philosophical arguments that allow us to suspect that there is a limit to our capacity to observe ourselves. The first is that each thought is unique and never repeats. In philosophy, the type/token distinction separates the types as a concept and an abstract object from the token, which is the realization, or instantiation, or occurrence of the type. One can think twice about the same dog, even in the same place and in the same light, but they are still two different thoughts. The second philosophical objection stems from a logical argument known as Leibniz's Law, which maintains that a subject is, at least in some way, unique and different from others. When an observer decodes another's mental states with maximum resolution, they do so from their own perspective, with their own nuances and overtones. In other words, the human mind has an impregnable sphere of privacy. Perhaps at some future time that sphere may become very small, but it cannot be eliminated altogether. If someone were to entirely access someone else's mental contents, then they would be that someone else. The two would meld. They would become one.

CHAPTER FIVE

The Brain is Constantly Transforming

What makes our brain more or less predisposed to change?

Is it true that it's much harder to learn things – like a new language or to play an instrument – when we are older? Why is it easy for some of us to learn music and so difficult for others? Why do we all learn to speak naturally, and yet many of us struggle with maths? Why is learning some things so arduous and others so simple?

In this chapter we will navigate into the history of learning, effort and virtue, mnemonic techniques, the drastic cerebral transformation when we learn to read, and our brain's capacity for change.

Virtue, oblivion, learning, and memory

Plato tells of a stroll in fifth-century BC Athens during which Socrates and Menon heatedly discussed virtue. Is it possible to learn it? If so, how? In the midst of the debate, Socrates presents a phenomenal argument: virtue cannot be learned. What's more, nothing can be learned. Each of us already possesses all knowledge. So learning actually means remembering.* This conjecture, so beautiful and bold, was

---

* In Latin, *cor* is literally the *heart* and is found in some Romance-language

implanted in different versions of Socratic teaching in thousands and thousands of classrooms across the world.

It's strange. The great master of antiquity was questioning the more intuitive version of education. Teaching is not transmitting knowledge but rather teachers help their students to express and evoke knowledge they already have. This argument is central to Socratic thought. According to him, at each birth, one of the many souls wandering about in the land of the gods descends to confine itself within the newborn body. Along the way it crosses the River Lethe, where it forgets everything it knew. It all begins with oblivion. The path of life, and of pedagogy, is a constant remembering of that which we forgot when crossing the Lethe.

Socrates proposed to Menon that even the most ignorant of slaves already knows the mysteries of virtue and the most sophisticated elements of maths and geometry. When Menon showed his incredulity, Socrates did something extraordinary, he suggested resolving the discussion in the realm of experiments.

The universals of human thought

Menon then called over one of his slaves, who became the unexpected protagonist of a formidable landmark in the history of education. Socrates drew a square in the sand and fired off a volley of questions. Just as mathematical works are a record of the most refined and elaborate Greek thought, Menon's slave's answers revealed the popular intuitions, the common sense, of the period.

In the first key passage of the dialogue, Socrates asks: 'How must I change the length of the sides so a square's area doubles?' Think of an answer quickly, make a hunch without getting into elaborate reflec-

---

words as 'passing through the heart', i.e. remembering, and in English shows up in *cordial* and *discourage*. In English we can know things *by heart*, and we find a similar metaphor in the etymology of *remember*, 'to pass through a body part'.

tion. That is probably what the slave did when he responded: 'I simply double the length of the sides.' Then Socrates proceeded to draw the new square in the sand and the slave discovered that it was made up of four squares identical to the original one.

The slave then discovered that doubling the side of the square quadrupled its area. And Socrates continued his question-and-answer game. Along the way, by responding on the basis of what he already knew, the slave expressed the geometrical principles that he intuited. And he was able to learn from his own errors and correct them.

Towards the end of the dialogue, Socrates drew a new square in the sand, whose side was the diagonal of the original square.

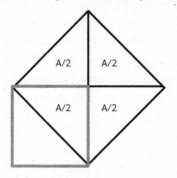

And then the slave could clearly see that it was made up of four triangles, while the original was made up of only two.

'Do you agree that this is the side of a square whose area is twice the original's?' asked Socrates.

To which the slave answered yes, thus sketching out the basis of the Pythagorean theorem, the quadratic relationship between the sides and the diagonal.

The dialogue concludes with the slave, who realizes, just by answering questions, the basis of one of the most highly valued theorems in Western culture.

'What do you think, Menon? Did the slave express any opinion that was not his own?' asked Socrates.

'No,' replied Menon.

The psychologist and educator Antonio Battro understood that this dialogue was the seed of an unparalleled experiment into whether there are intuitions that persist over centuries and millennia. I undertook this task with my graduate student Andrea Goldin, a biologist. We posed Socrates' questions to children, teenagers and adults and found that their responses, 2,500 years after the original dialogue, were almost identical. We are very similar to the Ancient Greeks,* we get the same things right and we make the same mistakes. This shows that there are ways of reasoning that are so deeply entrenched that they travel in time through cultures, changing little.

It doesn't matter – here – whether the Socratic dialogue actually took place or not. Perhaps it was merely a mental simulation by Socrates, or by Plato. However, we did show that it is plausible for the dialogue to have happened just as it was written. When faced with the same questions, people respond – millennia later – just as the slave did.

---

* You only have to watch the film *Troy* to notice an extraordinary resemblance between Achilles and Brad Pitt.

My motivation for doing this experiment was to investigate the history of human thought and examine the hypothesis that simple mathematical intuitions expressed in fifth-century Athens could be identical to those expressed by twenty-first-century students in South America or elsewhere in the world.

Andrea's motivation was different. Her drive was to understand how science can improve education – an ambition I learned by her side – and this led her to ask very different questions in the same experiment: was the dialogue really as effective as presumed? Is answering questions a good way to learn?

The illusion of discovery

In our experiment, Andrea proposed, once the dialogue was finished, to show each student a new square of a different colour and different size and ask them to use it to generate a new one with twice the area. It seemed to me that the task was too easy; it couldn't be exactly the same as what had been taught. So I suggested we test what they'd learned in a more demanding way. Could they extend the rule to new shapes; for example, a triangle? Could they generate a square whose area was half – instead of double – the original square?

Luckily Andrea stood her ground. As she had supposed, a large number of the participants – almost half, in fact – failed the simpler test. They couldn't replicate what they believed they had learned. What happened?

The first key to this mystery has already appeared in this book; the brain, in many cases, has information that it cannot express or evoke explicitly. It is like having a word on the tip of your tongue. So the first possibility is that this information was effectively acquired through the dialogue but not in a way that can be used and expressed.

An example in daily life can help us to understand the mechanisms in play. Someone is given a lift time after time to the same destination.

One day they have to take the wheel to follow the route they have been driven along thousands of times and find out they don't know where they're going. This doesn't mean that they didn't look at the route, or weren't paying attention. There is a process of consolidating knowledge that needs practice. This argument is central throughout the problem of learning; it is one thing to assimilate knowledge *per se* and another to assimilate it in order to be able to express it. A second example is the learning of technical skills, like playing the guitar. We watch the teacher, we see clearly how he or she articulates their fingers in order to make a chord, but, when it is our turn, we are unable to do the same thing.

The analysis of the Socratic dialogue shows that just as extensive practice is needed to learn procedures (playing an instrument, reading or riding a bike), it is also necessary for conceptual learning. But there is a crucial difference. In learning an instrument we recognize immediately that seeing is not enough to learn. Yet with conceptual learning both the teacher and the student feel that a well-sketched-out argument can be taken in without difficulty. That is an illusion. In order to learn concepts, meticulous practice is required, just as when learning to type.

Our further exploration of Menon's dialogue revealed a pedagogical disaster. The Socratic process turns out to be very gratifying for the teacher. The students' response seems highly successful. But when the class is put to the test, the result is not always so promising. My hypothesis is that this educational process sometimes fails for two reasons: the lack of practice in using the acquired knowledge and the focus of attention, which should not be placed on small fragments of facts that are already known but rather on how to combine them to produce new knowledge. We have already sketched out the first argument and we will explore it in further depth over the next few pages. A concise example of the second can be found in educational practice.

Beyond demographic, economic and social factors – which are, of course, decisive – there are countries in which mathematical teaching works better than in others. For example, in China students learn more than is expected – based on the GDP and other socioeconomic variables – and in the United States less. What explains this difference?

In the United States, teachers solving a complex multiplication like 173 × 75 on the blackboard usually ask the children things they already know: 'How much is 5 × 3?' And they all, in unison, answer: 'Fifteen.' It is gratifying because the entire class gets the right answer. But the trap lies in the fact that the children were not taught the only thing they didn't know, the path. Why start with 5 × 3 and then do 5 × 7, and not the other way around? How should they combine this information and how do they establish a plan for being able to resolve the other steps in the problem 173 × 75? This is the same error we found in the Socratic dialogue. Menon's slave would never have drawn the diagonal on his own. The big secret to solving this problem is not in realizing, once the diagonal has been drawn, how to count the four triangles. The key is in how to get the student to come up with the idea that the solution requires thinking of the diagonal. The pedagogical error lies in bringing the student's attention to fragments of the problem that had already been solved.

In China, on the other hand, in order to learn to multiply 173 × 75, the teacher asks: 'How do you think this is solved? Where do we start?' This, first of all, takes the students out of their comfort zone, inquiring about something they do not know. They have to establish how to break down this complex calculation into a series of steps: First multiply 5 × 3, and write the result, and then multiply 5 × 70, and so on . . . Secondly, it leads them to make an effort and, eventually, to make mistakes. The two teaching methods coincide in that they are based on questions. But one asks about already known fragments and the other about the path that unites the fragments.

Learning through scaffolding

In our investigation into the contemporary responses to Menon's dialogue we found something odd. Those who followed the dialogue to the letter learned less. On the other hand, those who skipped over some questions learned more. The odd thing is that more teaching – more of the dialogue – favoured less learning. How can we resolve this enigma?

We found the answer in a research programme carried out by the psychologist and educator Danielle McNamara in order to decipher a text's legibility. Her project, vastly influential in the worlds of academia and educational practice, shows that the most pertinent variables are not the ones you would expect, such as attention, intelligence and effort. The most decisive, in fact, was what the reader already knew about the subject before starting.

This led us to a very different reasoning from the one any of us would have naturally sketched out in the classroom; learning doesn't fail because of distraction or lack of attention. In fact, students with almost no prior knowledge can follow the dialogue with great concentration, but their attention is focused on each step, on the trees and not the forest; while those students whose knowledge already brings them close to the solution will not need so much concentration to follow the dialogue.

So Andrea and I sketched out a seemingly paradoxical hypothesis: those who pay more attention learn less. In order to test it out we created a pioneering experiment, the first simultaneous recording of cerebral activity while one person was teaching and another was learning.

The results were conclusive. Those who learned less activated their prefrontal cortex more, which is to say, they made more effort. To such an extent that, by measuring cerebral activity during the dialogue, we could predict whether a student would later pass the exam.

Of course, it is not always true that paying more attention means learning less. With equal prior knowledge, those who pay more attention achieve greater results. But in this dialogue – and so many others in school – it turns out that effort is inversely related to prior knowledge. Those with less knowledge follow the dialogue step by step, in detail. Yet those who are able to skip over whole parts can do so because they already know many of the fragments. The path is well learned only when one can follow it, without needing to stop at every step.

This idea is closely linked to the concept introduced in the 1920s by the great Russian psychologist Lev Vygotsky of the *zone of proximal development*, which made such an impact on pedagogy. Vygotsky argued that there must be a reasonable distance between what students can do for themselves and what a mentor demands of them. Later in the book we will revisit this idea when looking at how to lessen the gap between teachers and students by having the children themselves act as mentors. But at this point I want to leap through another window that was opened in the minute analysis of the Socratic dialogue: learning, effort and leaving one's comfort zone.

Effort and talent

We intuit that the few people who learn to play the guitar like a rock star such as Prince* do so based on a certain mix of biological and social factors. But to understand how these elements interact with each other and, above all, how to use that knowledge better to learn and teach, we need to divide this general concept into smaller sections.

The idea that genetic factors determine the maximum skill that each of us can achieve is very deep-seated. In other words, anyone can learn music or football to a certain level, but only a few virtuosos can reach the level of João Gilberto** or Lionel Messi. Great talents are born, not made. They were touched with a magic wand, they have a *gift*.

This idea that we all go through a similar educational trajectory, but the ceiling depends on a biological predisposition, was coined

---

* For example, Prince.

** In *Pra ninguém*, Caetano Veloso lists the pieces of music that move him the most. And then he says: 'But better than all of them is silence. And better than silence, only João.'

and sketched in 1869 by Francis Galton, one of the most versatile and prolific British scientists. The clearest example appears when the predisposition is a body trait. For example, becoming a professional basketball player is much more likely if you are tall. It is hard to become a great tenor without having been born with the proper vocal apparatus.

Galton's idea is simple and intuitive but doesn't coincide with reality. When investigating in detail how great experts learned what they know, and avoiding the temptation to draw general conclusions based more on myth, it turns out that the first two premises of Galton's argument are wrong. The upper limit of learning is not so genetically based, nor is the path towards that upper limit so independent of genetics. Genetics are involved in both parts, but not decisively in either.

Ways of learning

The great neurologist Larry Squire sketched a taxonomy that divides learning into two large categories. *Declarative* learning is conscious and can be explained in words. A good example of this is learning the rules of a game; once the instructions are learned, they can be taught (declared) to a new player. *Nondeclarative* learning includes skills and habits that are usually achieved without the learner being aware of the process. These are types of knowledge that would be difficult to make explicit in the form of language, such as by explaining them to someone else.

The more implicit ways of learning are, in fact, so unconscious that we don't even recognize that there was something to learn: for example, learning to see. We can easily identify that a face expresses an emotion but we are unable to *declare* this knowledge in order to make machines that can emulate this process. Our ability to see is innate in most people. So much so that the inverse of this naturalness of the gaze has poetic strength. The Uruguayan author Eduardo Galeano wrote: 'And the sea was so immense, so brilliant, that the boy was struck dumb by its beauty. And when he finally managed to speak, trembling, stuttering, he asked his father: "Help me see!"' A similar thing happens when learning to walk or keep our balance. These faculties are so well-incorporated that it seems they've always been there, that we never had to learn them.

These two categories are useful when exploring the vast space of learning. However, it is equally important to understand that they are inevitably abstractions and exaggerations; almost all learning in real life is part declarative and part implicit.

For example, learning to walk is an implicit and procedural form of learning, it doesn't require instructions or explanations, and it's learned slowly and after a lot of practice. Yet there are many aspects which can be consciously controlled. The same thing happens with breathing, which is fundamentally an unconscious process. It would not be sensible to delegate to each of our distracted free wills something that would be fatal if forgotten. But, to a certain point, we can control our breathing consciously, its rhythm, volume, flow. And it is breathing, the bodily function that spans the conscious and the unconscious, that is used as a universal bridge in meditative practices and other exercises to learn to direct one's consciousness to new places.

Establishing this bridge between the implicit and the declarative turns out to be, as we will see, a key variable in every form of learning.

The OK threshold

A fundamental concept for understanding how much we can improve is called the *OK threshold*, the level at which everything feels fine. People engaged in learning to type, for example, begin by searching out each letter with their eyes, exerting great effort and concentration. Like Menon's slave, they pay attention to each step. But later it seems as if their fingers have a life of their own. When we touch-type, our brains are somewhere else, reflecting on the text, talking to someone else or daydreaming. What's curious is that once we have reached this level of ability, despite typing for hours and hours, we no longer improve. In other words, the learning curve grows until it reaches a value where it stabilizes. Most people reach speeds close to sixty words per minute. But, of course, this value is not the same for everyone; the world record is held by Stella Pajunas, who managed to type at the extraordinary pace of 216 words per minute.

This example seems to confirm Galton's argument; he maintained that each of us reaches our own inherent ceiling. Yet by doing methodical, sustained exercises to increase our speed, all of us can improve substantially. What happens is that we stagnate very far from our maximum performance, at a point at which we benefit from what we've learned but we do not generate further learning, a comfort zone in which we find a tacit balance between the desire to improve and the effort that would require. This point is the *OK threshold*.

The history of human virtue

What happens in the example of typing speed occurs with almost everything we learn in life. One example that most of us experience is reading. After years of intense effort at school, many of us achieve reading quickly and with little effort. We read more and more books, without increasing our reading speed. Yet if any of us revisited a methodical, sustained process, and devoted time and effort, we could significantly increase our speed without losing comprehension along the way.

The narrative of learning in each of our life cycles is replicated in the history of culture and sport. In the early twentieth century, the fastest runners of the times achieved the extraordinary feat of running a marathon in two and a half hours. In the early twenty-first century, this time isn't enough to qualify for the Olympics. This is not limited to sports, of course. Some compositions by Tchaikovsky were technically so difficult that in his day they were never played. The violinists of the period thought that they were impossible. Today they are still considered challenging but there are many violinists who can play them.

Why is it that we can now achieve feats that years earlier were impossible? Is it that, as Galton's hypothesis suggests, our constitution – our genes – changes? Of course not. Human genetics, over those seventy years, has remained essentially the same. Is it because technology has radically changed? Again, the answer is no. Perhaps that would be a valid argument for some disciplines, but a marathon runner with trainers from a hundred years ago – and even barefoot – could today achieve times that were once impossible. Likewise, a contemporary violinist could now play Tchaikovsky's works with period instruments.

This deals a fatal blow to Galton's argument. The limits of human performance are not genetic. Violinists today manage to play those pieces because they can devote more hours to their practice, because the point at which they feel the goal is accomplished has changed, and because they have better training procedures. This is good news; it means that we can build on these examples to attain goals that today are inconceivable.

Fighting spirit and talent: Galton's two errors

When we judge athletes we usually separate their competitive spirit from their talent as if they were two different substrata. There are the Roger Federers of the world – who have talent – and the Rafael Nadals – who are mainly driven by intense competitive spirit to the extent that they leave their bodies and souls on the field. A typical observer views those with innate talent with a distant respect that denotes admiration for a gift, a competitor's divine privilege. Fighting spirit, on the other hand, feels more human because it is associated with will and the feeling that we could all achieve it. This is Galton's hunch: the gift, the ceiling of talent, is innate, and fighting spirit, the path to advancement through learning, is available to all of us. Both of these conjectures, however, are wrong.

In fact, the ability to *give one's all* out on the pitch is perhaps one of the elements most determined by genetic makeup. It is, in fact, an element of temperament, which refers to a vast term that defines personality traits including emotiveness and sensitivity, sociability, persistence and focus. In the mid-twentieth century, an American children's psychiatrist, Stella Chess, and her husband, Alexander Thomas, began a *tour de force* study that would be a landmark in the science of personality. As in a film by Richard Linklater, they meticulously followed the development of hundreds of children from different families, from the day they were born through adulthood. They measured nine traits of their temperaments:

(1) Their activity levels and types.
(2) The degree of regularity in their diet, and sleeping and waking habits.
(3) Their willingness to try something new.
(4) Their adaptability to changes in the environment.
(5) Their sensitivity.
(6) The intensity and energy level of their responses.
(7) Their general mood – happy, weepy, pleasant, nasty or friendly.
(8) Their degree of distraction.
(9) Their persistence.

They found that while these traits were not immutable, they at least persisted to a striking extent throughout their development. And, what's more, they were expressed clearly and precisely in the first days of life. Over the last fifty years, this foundational study by Chess and Thomas has been continued with a multitude of variations. The conclusion is always the same: a significant portion of the variance – between 20 and 60 per cent – in temperament is explained by the genetic package we are born with.

If genes explain more or less half of our temperament, the other half is explained by the environment and social petri dish in which we develop. But which specific elements of our environment? Of almost all cognitive variables, the most decisive factor is the home a child grows up in. Siblings are similar not only because they carry similar genes but also because they develop on the same playing field. But there are exceptions. Different studies on adoptions and twins show that the home contributes very little to the development of some aspects of temperament.

Searching for the nature of human altruism, an Austrian behavioural economist, Ernst Fehr, has shown this quite conclusively for one of the foundational traits of temperament: the predisposition to sharing. When children choose between keeping two toys or sharing them equally with a friend – throughout diverse cultures, different continents and different socioeconomic strata – the younger sibling is usually less predisposed to share. In retrospect, this seems natural; the younger one was raised according to 'if you don't ask you don't get'. When the younger siblings get something, they keep it for themselves in their jungle filled with older predators. All parents with more than one child recognize that the anxiety, fragility and above all ignorance in which a first child is raised are not repeated. As a result of this, some social aspects of young children, such as their disposition to share, do not depend so much on the home experience but are rather learned on other playing fields of life.

Our excursion to the science of temperament sheds light on why Galton's intuition – which persists today as a very popular myth – is wrong. It certainly feels as if we could all potentially attain the ability to give one's all – as opposed to talent, which feels like a natural gift that only very few have. But in the list of temperament traits that vary little through our lives, we find the main ingredients for giving one's all: the intensity and the energy of the responses, the general mood, the degree of distraction, the capacity to persist and the intensity of basal activity. And with this we can understand why the capacity of giving one's all is an ability that varies widely across individuals and is quite difficult to change.

This explanation is mostly based on the work of Stella Chess and Alexander Thomas, who have meticulously observed the persistence and malleability of different personality traits that make us what we are. We still need to understand what specific aspects of our biological constitution, of our genes and of our brains, regulate the ability to give one's all. The answer to this question is, in my opinion, far from being complete. But we will see later in this chapter that it is closely related to aspects of the motivation and reward system that define temperament and are a gateway to learning.

Now we must topple the opposite myth. What we perceive as talent is not an innate gift but rather, almost always, the fruit born of hard work. Let's use an emblematic case to defend this argument: perfect pitch, the ability to recognize or produce a musical note without any point of reference. Perfect pitch is one of the most widely recognized cases of a gift of talent. Somebody with this aptitude is usually considered a musical genius, so out of the ordinary as to be viewed as some sort of mutant, like an X-Man of music, gifted with a genetic package that gives them this unusual virtue. Once again, a lovely idea . . . but erroneous, a myth.

Perfect pitch can be trained, and almost everyone can achieve it. In fact, most children have *almost* perfect pitch, but without practice it atrophies. And children who begin training at a conservatory at a young age have a very high incidence of perfect pitch. Once again, this is not due to genius but to hard work. Diana Deutsch, one of the finest researchers on music and the brain, made an extraordinary discovery: people who live in China and Vietnam have a much higher predisposition to perfect pitch. What is the origin of this peculiarity? It turns out that in Mandarin and Cantonese, as well as in Vietnamese, words change their meaning based on tone. So, for example, in Mandarin the sound 'ma' pronounced in different tones can mean mother or horse and, if that weren't confusing enough, it also means marijuana. So tone has an absolute value – as much as the musical note F is different from D or G – and there is a higher motivation for learning this relationship between a particular tone and the meaning it represents in China – to distinguish mother from horse, for example – but not as much in other parts of the world. Therefore, the motivation and pressure exerted by the language extend to music in something that ends up being much less sophisticated and less revealing of genetics and geniuses than it seems.

The fluorescent carrot

While I was completing my doctorate in New York, a group of friends and I played an absurd game. We tried to control the temperature of our fingertips, which isn't the most exciting achievement in the world, but does demonstrate an important principle, namely that we can voluntarily regulate certain aspects of our physiology that are seemingly inaccessible. We were, in the fantasy of those moments, students of Charles Xavier at the school of young mutants.

With a thermometer on my fingertip, I observed that the temperature fluctuated between 31 and 36 degrees Celsius. Then I tried to raise that temperature. Sometimes I was successful and the temperature of my fingertip became raised and sometimes it didn't. These variations were spontaneous and random and proved that in spite of my wish to do so I couldn't control them. However, after two or three days of practice something astonishing happened. I managed to manipulate the thermometer at will, although imprecisely. Two days later, I had perfect control. I could control the temperature of my fingertip just by using my thoughts. Anyone can do it. This learning process is mysterious because it is not declarative. It is likely that I learned to relax my hand, thus changing the blood flow and controlling the temperature. But I couldn't – and still can't – precisely explain in words what exactly it was that I learned.

This innocent game reveals a fundamental concept for many of the brain's learning mechanisms. For example, when trying to move the arm for the first time to reach something, an infant explores a large repertoire of neuronal commands. Some, coincidentally, turn out to be effective. And here is the first key point: in order to select efficient commands one must visualize their consequences. Later this mechanism becomes more refined and the baby has no need to rehearse all of the neuronal commands. For the ones that have been selected, the brain generates an expectation of success, which allows learners to simulate the consequences of their actions without having to carry them out, like football players who don't run after the ball because they know they can't reach it.

And therein lies the second key point in learning, known as prediction error, which we have already touched on in Chapter 2. The brain calculates the difference between what is expected and what is in fact achieved. This algorithm allows us to refine the motor programme and, with that, achieve a much more precise control over our actions. That is how we learn to play tennis or an instrument. This learning mechanism is so efficient that it became common currency in the world of automatons and artificial intelligence. A drone literally learns to fly, and a robot to play ping-pong, using this procedure, which is as simple as it is effective.

In the same way, we can learn to control all sorts of devices with our thoughts. In a not-so-distant future, the projection of this principle will generate a landmark in the history of humanity. The body might no longer be necessary as an intermediary. It would be enough to want to call someone for a device to decode the gesture and carry it out without hands or voices, without a mediating body. In the same way, we can extend the sensory landscape. The human eye is not sensitive to colours beyond violet, but there is no essential limit to this. Bees, for example, get to see in the ultraviolet world. We can use photographic techniques to mimic that world, but the resulting colours are only approximations of what a bee might see. Bats and dolphins also hear sounds that are inaudible to our ears. Nothing stands in the way of us someday connecting electronic sensors capable of detecting this vast portion of the universe that today is opaque to our senses. We can also impregnate ourselves with new senses. For example, having a compass directly connected to the brain in order to allow us to *feel* the north in the same way we feel the cold. The mechanism for achieving it is essentially the same as the one I described in the innocuous game of the fingertip temperature. The only difference is the technology.

This learning procedure requires being able to visualize the consequences of each neuronal instruction. So by increasing the range of things we visualize, we also widen the number we learn to control. Not only in terms of external devices but also in our inner world, in our own body.

Controlling the temperature of your fingertip with your will is undoubtedly a trivial example of this principle, but it sets an extraordinary precedent. Is it possible for us to train the brain to control aspects of our bodies that seem completely detached from our consciousness and from the realm of things we can exert our will on? What if we could visualize the state of our immune system? What if we could visualize states of euphoria, happiness or love?

I would venture to say that we will be able to improve our health when we manage to visualize aspects of our physiology that are currently invisible to us. This already is the case in a few specific areas. For example, it is now possible to visualize the pattern of cerebral activity that corresponds to a state of chronic pain and, based on this visualization, control and lessen it. This could be taken much further, and we would be able to regulate our defence system in order to overcome diseases that now seem insurmountable. Research could be focused on this fertile territory for what today seems like miraculous healing but in the future could be visualized and made standard.

The geniuses of the future

The myth of genetic talent is based on rare cases and exceptions, on stories and photos that show precocious geniuses with their innocent youthful faces rubbing elbows with the big names of the world's elite. The psychologists William Chase and Herbert Simon toppled this myth by closely investigating the progression of the great chess geniuses. None of them achieved an exceptionally high level of skill before having completed 10,000 hours of training. What was perceived as precocious genius was shown to be based, in fact, upon intensive and specialized training from a very young age.

The vicious circle functions more or less like that: the parents of little X convince themselves that their offspring is a violin virtuoso and they give the child the confidence and motivation to practise, and, therefore, X improves greatly, to the point of seeming talented. Acting as if someone has talent is an effective way to get them actually to have it. It seems to be a self-fulfilling prophecy. But it is much more subtle than the mere psychological configuration of 'I think, therefore I am.' The prophecy produces a series of processes that catalyse the more difficult aspect of learning: tolerating the tedium of the effort involved in deliberate practice.

All this comes up hard against the more extreme exceptions. What do we do with what seems obvious? For example, that Messi was already an indisputable football genius from a very young age. How do we match up the detailed analysis of development by experts with what our intuition tells us?

Firstly, the argument of effort does not rule out the existence of a certain innate condition.* But, in addition, believing that he wasn't an expert at eight years old is the beginning of the wrongheaded thinking. At that age, Messi already had more football experience than most people on the planet. The second consideration is that there are hundreds – thousands – of children who do extraordinary things with a ball. But only one of them grew up to be Leo Messi. The error is in supposing that we can predict which children will be the geniuses of the future. The psychologist Anders Ericsson, with careful monitoring of the education of virtuosos in various disciplines, proved that it is almost impossible to predict how much an individual might achieve based on performance in the early stages. This final blow to what we think we know about nurturing talent and effort is very revealing.

The expert and the novice use completely different systems of

---

* Such as Manu Ginóbili, with his height for basketball; or like X, with that name, for the violin.

resolution and cerebral circuits, as we will see. Learning how to do something skilfully is not about improving the cerebral machinery with which we would originally resolve it. The solution is much more radical: replacing it completely for one with different mechanisms and idiosyncrasies. This idea was first hinted at by the celebrated study that Chase and Simon made of expert chess players.

A circus trick that some great chess players practise from time to time is playing games with their eyes closed. Some are capable of extraordinary feats. Miguel Najdorf played on forty-five different boards simultaneously, with his eyes blindfolded. He won thirty-nine games, reached a draw in four, and lost two, breaking the world record for simultaneous games.

In 1939 Najdorf had travelled to Argentina to participate in a Chess Olympiad, representing Poland. Najdorf was Jewish, and as the Second World War had broken out while he was away, he decided not to come back to Europe. His wife, son, parents and four brothers died in a concentration camp. In 1972, Najdorf explained the personal reasons behind his remarkable deed of playing forty-five boards: 'I didn't do it for the fun of pulling it off. I had the hope that this news would reach Germany, Poland, and Russia, and that one of my family members would read it and get in touch with me.' But no one did. The greatest human feats are, in the end, a struggle against loneliness.*

There were 1,440 pieces in play on 45 boards; 90 kings, 720 pawns. Najdorf followed them all in unison to lead his 45 armies, split between black and white, with his eyes covered. Of course, he must have had an extraordinary memory, and have been a very special, unique person with a true gift to be able to do this. Or was he?

A grandmaster, just by looking at a diagram of a chess match for a

---

* Najdorf's grandson told me that Don Miguel was only able to find one of his cousins. It was by chance. On the subway in New York they recognized the similarities between them, started a conversation and discovered they were related.

few seconds, can reproduce it perfectly. Without any effort, as if their hands were working on their own steam, the chess master can place the pieces exactly where the diagram indicates they should be. Yet a person who doesn't know chess, when faced with this same exercise, would barely remember the position of four or five pieces. It would indeed seem that chess players have much better memories. But that is not the case.

Chase and Simon proved this by using diagrams with pieces spread out randomly on the table. Under those conditions, the masters only remembered, like everyone else, a few pieces. Chess players do not have extraordinary memories but rather the ability developed through practice to create a narrative – visual or spoken – for an abstract problem. This discovery does not only hold true for chess, but for any other form of human knowledge. For example, anyone can remember a Beatles song but will have difficulty recalling a sequence made up of the same words but presented randomly. *Now try to remember this same sentence that is long but not complex.* And this one: *sentence but same long complex try now not remember this that.* The song is easy to remember because the text and the music have a narrative. We do not remember it word by word but rather we remember the path the words comprise.

Heirs to Socrates and Menon, Chase and Simon hit upon the key to establishing the path to virtue and knowledge. And the secret, as we will see, consists in recycling old brain circuits to adapt themselves to new functions.

Memory palace

Mnemonic skill is often confused with genius. Someone who juggles with their hands is skilful, but someone who does so with their memory seems to be a genius. And yet, they are not really all that different. We learn to develop a prodigious memory the same way we learn to play tennis, with the recipe we've already discussed: practice, effort, motivation and visualization.

When books were rare objects, all stories were disseminated orally. To keep a story from vanishing, people had to use their brains as memory repositories. So, out of necessity, many were seasoned at learning by rote. The most popular mnemonic technique, called 'the memory palace', was created in that period. It is attributed to Simonides, the Greek lyric poet from the island of Ceos. The story goes that Simonides had the luck to be the only survivor of the collapse of a palace in Thessalia. The bodies were all mutilated, so it was almost impossible to recognize them and bury them properly. All they had was Simonides' memory. And he realized, to his surprise, that he could vividly recall the exact place where each of the guests had been sitting when the building collapsed. As a result of this tragedy he had discovered a fantastic technique, the memory palace. He understood that he could remember any arbitrary list of objects if he visualized them in his palace. This was, in fact, the beginning of modern mnemotechnics.

With his palace, Simonides identified an idiosyncratic trait of human memory. The technique works because we all have a fabulous spatial memory. To prove this, you only need to think of how many maps and routes (through towns or people's homes, of buses, through cities and buildings) you can recall with no effort whatsoever. This original seed of a discovery came to fruition in 2014, when John O'Keefe and the Norwegian couple May-Britt and Edvard Moser won the Nobel Prize in Medicine for finding a system of coordinates in the hippocampus that articulates this formidable spatial memory. It is an age-old system that is even more refined in small rodents – who are extraordinary navigators – than in our mastodon minds. Situating ourselves in space has always been necessary. Unlike remembering countries' capitals, numbers and other things, which our brains *never evolved to do.*

Here we see an important idea. An ideal way for us to adapt to new cultural needs is by recycling brain structures that evolved in other contexts to fulfil other functions. The memory palace is a very paradigmatic example. We all struggle to remember numbers, names or shopping lists. But we can easily recall hundreds of streets, the nooks and crannies of our childhood homes, or the houses of our friends when we were growing up. The secret of the memory palace is establishing a bridge between those two worlds, what we want to remember, but find it difficult to, and space, where our memory is right at home.

Read this list and take thirty seconds to try to remember it: napkin, telephone, horseshoe, cheese, tie, rain, canoe, anthill, ruler, tea, pumpkin, thumb, elephant, barbecue, accordion.

Now close your eyes and try to repeat it in the same order. It seems difficult, almost impossible. Yet someone who has built their palace – which requires a few hours of work – can easily remember a list like this. The palace can be open or closed, an apartment building or a house; then what you have to do is go through each room and place, one by one, each of the objects on the list. You have to do more than just name them. In each room you have to create a vivid image of the object in that place. The image must be emotionally powerful, perhaps sexual, violent or scatological. The unusual mental stroll, in which we peek into each room and see the most bizarre images featuring those objects in our own palace, will persist in our memories much more than the words.

So a prodigious memory is based on finding good images for those objects we wish to remember. The task of memorizing is somewhere between architecture, design and photography, all creative tasks. Memory, which we tend to perceive as a rigid and passive aspect of our thought, is actually a creative exercise.

In short, improving our memories doesn't mean increasing the space in the drawer where the memories are kept. The substratum of memory is not a muscle that grows with exercise and is strengthened. When technology made it possible, Eleanor Maguire confirmed this premise by investigating the very factory of memories. She discovered that the brains of the great champions of memory are anatomically indistinguishable from anyone else's. Nor were those champions more *intelligent* or better at remembering things outside of the realm they had studied, just like virtuoso chess players. The only difference was that the prodigious memorizers use the spatial structures of their memories. They have managed to recycle their spatial maps in order to remember arbitrary objects.

The morphology of form

One of the most spectacular cerebral transformations occurs when we learn to see. This happens so early in our lives that we do not have any memory of how we perceived the world before seeing. From a stream of light, our visual system manages to identify shapes and emotions in a tiny fraction of a second, and what is even more extraordinary is that it happens without any sort of effort or conscious realization that something must be done. But converting light into shapes is so difficult that we have yet to create machines that can do it. Robots go into outer space, play chess better than the greatest masters, and fly aeroplanes, but they cannot see.

To understand how the brain pulls off such a feat, we must find its limits, see exactly where it fails. To do that, we will look at a simple yet eloquent example. When we are trying to think about how we see, an image is definitely worth more than a thousand words.

The two objects in the following figure are very similar. And both, of course, are very easy to recognize. But when they are submerged in a sea of dotted lines something quite extraordinary happens. The visual brain works in two completely different ways. It is impossible not to see the object on the right; it's as if it were another colour and was literally popping out. Something different happens with the object on the left. We see the dotted lines that make up the snake only with much effort, and our perception is unstable; when we are focused on one part, the rest vanishes and blends into the texture.

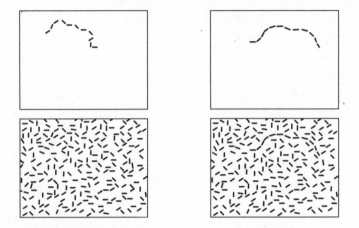

We can think of the object we see easily as a melody in which the notes follow each other harmonically and are naturally perceived as a whole, while the other object is more like random notes. Just as with music, the visual system has rules that define how we organize an image and that dictate how we perceive and what we remember. When an object is grouped in a natural, integrated way that does not require much effort, it is said to be *gestaltian*, named for the group of psychologists who, in the early twentieth century, discovered the rules by which the visual system constructs shapes. These rules, like those of language, are learned.

Let's look at how this works in the brain. Can we train and modify the brain to detect any object almost instantaneously and automatically? In the process of answering this question, we will sketch a theory of human learning.

A monster with slow processors

Most silicon computers today work with just a few processors. These computers calculate very quickly, but can only respond to one thing at a time. Our brains, on the other hand, are *parallel* machines – to a massive degree; which is to say, they are simultaneously making millions and millions of calculations. Perhaps this is one of the most distinctive aspects of the human brain and, to a large extent, it allows us quickly and effectively to solve problems that we are still unable to delegate to even the most powerful computers. This is one of the areas of most intense effort in computer science and yet the attempt to develop massively parallel computers has produced only elusive results. These researchers are presented with two fundamental difficulties: the first is simply to find a way to produce that number of processors economically; and the second is to get them all to share information.

In a parallel computer, each processor does its job. But the result of all that collective work has to be coordinated. One of the most mysterious aspects of the brain is how it manages to unite all the information processed in parallel. This is profoundly linked to consciousness. Which is why, if we understand how the brain brings together the information it calculates on a massive scale, we will be much closer to revealing the mechanism of consciousness. And we will have discovered the processes of how to learn.

The secret of virtuosity is in being able to recycle this parallel machinery and adapt it to new functions. The great mathematician *sees* mathematics. The chess master *sees* chess. And this happens because the visual cortex is the most extraordinary parallel machine known to humankind.

The visual system is comprised of superimposed maps. For example, the brain has a map that is devoted to codifying colour. In a region called V4,* modules of approximately a millimetre in size, called *globs*, are formed, and each one identifies different subtleties of colour in a very precise region of the image.

The big advantage to this system is that recognizing something does not require sequentially sweeping over it point by point. This turns out to be particularly important in the brain. It takes a long time

---

* The visual areas are named – to make it simple – with the letter V and a number that is a measurement of their place in the computing hierarchy, which means that V4 is one of the first stages of the more than sixty areas of visual processing.

for a neuron to load and route information from one to the next, which means that the brain can process between three and fifteen computing cycles in a second. That is nothing compared to the billions of cycles per second of the tiny processor in a mobile phone.

The brain solves the intrinsic slowness of its biological fabric with an almost infinite army of neural circuits.* So the conclusion is simple, and as I will argue in the next few pages will be the key to the enigma of learning: any function that can be resolved in the parallel structures (maps) of the brain will be done effectively and efficiently. It will also be perceived as automatic. On the other hand, functions that use the brain's sequential cycle are carried out slowly and are perceived with great effort and fully consciously. Learning in the brain is, to a great extent, *parallelizing*.

The repertoire of visual maps includes movement, colour, contrast and direction. Some maps identify more sophisticated objects as two contiguous circles. In other words, eyes that watch us. That is what produces the strange sensation of turning your head quickly towards someone watching you. How did you know they were watching you before turning your gaze? The reason is that the brain is exploring the possibility that someone is watching us throughout the entire space and in parallel, often without any conscious register of it. The brain detects a different attribute in one of its maps, and generates a signal that communicates with the attention and motor control system of the parietal cortex as if** saying: 'Turn your eyes over there because something important is happening.' These maps are like factory settings of a range of innate skills. They are efficient and at the same time fulfil a very specific task. But they can be modified, combined and rewritten. And in this lies the key to learning.

---

* 'They got the guns, we've got the numbers,' sings Jim Morrison in 'Five to One'.

** The 'as if' here is literal. The visual cortex doesn't speak in English with the parietal cortex. But these metaphors help us to understand how certain mechanisms work, as long as they are not too exaggerated or distracting.

Our inner cartographers

The cerebral cortex is organized into columns of neurons, and each one carries out a specific function. That was discovered by David Hubel and Torsten Wiesel, earning them the Nobel Prize in Physiology. When studying how those maps developed they found that there were *critical periods*. The visual maps have a natural programme of genetic development but they need visual experience to consolidate. Like a river that needs water running through it in order to maintain its shape.

The retina, especially in the first phases of development, generates spontaneous activity, stimulating itself in complete darkness. The brain recognizes this activity as light, without differentiating whether it comes from outside the body or not. Therefore the development based on activity begins before we open our eyes. Cats, for example, are born with their eyes closed. They are actually training their visual system with *inner light*. In cats, humans and other mammals, these maps develop in early infancy and are already consolidated after a few months. Hubel and Wiesel's discovery converges with another myth: that learning certain things as an adult is an impossible mission. We are going to revisit this idea and offer some moderate optimism: learning later in life is much more plausible than we imagine but requires much time and effort – the same amount we devote to such tasks in our early infancy, although we have now forgotten that. After all, babies and kids spend hours, days, months and years of their lives learning to speak, walk and read. What adult puts everything else to one side to devote all their time and effort to learning something new?

In retrospect, some of this is obvious. Radiologists, as adults, learn to *see* x-rays. After much work they can easily identify oddities that no one else can see. It is the clear result of a transformation in their adult visual cortex. In fact, for radiologists, such detection is quick, automatic and almost emotional, as when we have a visceral response of annoyance when we see 'grammatical' errors. What happens in the brain that can so radically transform our way of perceiving and thinking?

Fluorescent triangles

Science has some curious repetitions. Those who come up with extraordinary, paradigmatic ideas are often the very same ones who later topple them. Torsten Wiesel, after establishing the dogma of the critical periods, got together with Charles Gilbert, one of his students at Harvard, to prove just the opposite, that the visual cortex continues to reorganize itself even in full adulthood.

When I arrived in Gilbert and Wiesel's laboratory – by that point it had moved to New York – to begin my doctorate, the myth had already been turned on its head. The question was no longer whether the adult brain could learn, but how exactly it did. What happens in the brain when we become experts in something?

We devised an experiment to be able to carefully investigate this question in the laboratory. This required making certain concessions in a process of simplification. So instead of using expert radiologists, we created experts in triangles. Something with very little merit as a skill or profession but which has the laboratory advantage of being a simple way to simulate a learn-ing process.

We showed a group of people an image filled with shapes that after 200 milliseconds disappeared in a flash. They had to find a triangle in that mess. They looked at us like we were crazy. It was impossible. They simply hadn't had enough time to see it.

We knew that if the test had been finding a red triangle among many blue ones, everyone would have solved it easily. And we know why. We have a parallel system that, in eighty milliseconds, can sweep the space in unison in order to solve a difference of colour, but our visual cortices do not have a similar map devoted to identifying triangles. Can we develop that ability? If so, we would be opening a window on to how we learn.

Over hundreds of attempts, many of the participants were frustrated to find they saw nothing. But after hours and hours of repeating this boring task, something magical happened: the triangle began to glow, as if it were a different colour, as if it were impossible not to see it. So we know that with much effort we can see something that previously seemed impossible. And that it can be done as an adult. The big advantage to this experiment is that it allowed us to study what happens in the brain as we learn.

The parallel brain and the serial brain

The cerebral cortex is organized into two large systems. One is the dorsal, which – if your head is looking upwards – continues along the back of the body, and the ventral, which corresponds to the belly. In functional terms, this division is much more pertinent than the more popular division of the hemispheres. The dorsal part includes the parietal and frontal cortex, which have to do with consciousness, with cerebral activity that deals with action, and with a slow, sequential working of the brain. The ventral part of the cerebral cortex is associated with automatic and generally unconscious functions, and corresponds to a rapid, parallel way of working.

We found two fundamental differences in the cerebral activity of the *triangle experts*. Their primary visual cortex – in the ventral system – activated much more when they saw triangles than when they saw other shapes they hadn't been trained to identify. And at the same time their frontal and parietal cortices deactivated. This explains why for them seeing triangles no longer entailed effort. This is not specific to triangles. A similar transformation is observed when a person trains to recognize something (for example, musicians learning to read scores, a gardener learning to recognize a parasite on a plant, or coaches who realize in a matter of seconds that their team is flailing out on the field).

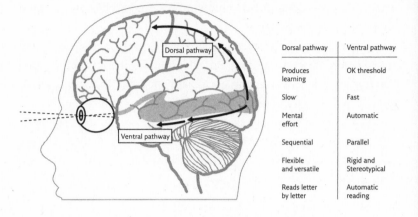

Dorsal pathway	Ventral pathway
Produces learning	OK threshold
Slow	Fast
Mental effort	Automatic
Sequential	Parallel
Flexible and versatile	Rigid and Stereotypical
Reads letter by letter	Automatic reading

Learning: a bridge between two pathways in the brain

The cortex is organized into the dorsal system and the ventral system. Learning consists of a process of transferring from one system to the other. When we learn to read, the slow, effortful system that works 'letter by letter' (dorsal system) is replaced by the other, which is capable of detecting entire words without effort and much faster (ventral system). But when the conditions are not favourable for the ventral system (for example, if the letters are written vertically) we go back to using the dorsal, which is slow and serial but has the flexibility to adapt to different circumstances. In many cases, learning means freeing the dorsal system to automatize a process so we can devote our attention and mental effort to other matters.

The repertoire of functions: learning is compiling

The brain has a series of maps in the ventral cortex that allow it to carry out some functions rapidly and efficiently. The parietal cortex allows for the combination of each of those maps, but through a slow process that requires effort.

However, the human brain has the ability to change its repertoire of automatic operations. After thousands and thousands of rehearsals, a new function can be added to the ventral cortex. We can think of this as a process of outsourcing, as if the conscious brain were delegating this function to the ventral cortex. Conscious resources, requiring mental effort and limited to the capacities of the frontal and parietal cortices, could be devoted to other tasks. This is a key to learning how to read that is enormously relevant to educational practice. Expert readers, who read cover to cover with no effort, delegate the reading; readers who are learning don't, and their conscious mind is fully occupied by the task.

The process of automatization is tangible in the example of arithmetic. When children first learn to add 3 + 4, they count on their fingers, making the parietal cortex work hard. But at some point in the learning 'three plus four is seven' almost becomes a poem. Their brains are no longer moving imaginary objects or real fingers one by one, but going to a memorized chart. The addition has been outsourced. Then a new phase starts. Those same children begin to solve 4 × 3 in the same slow, laboured way – using the parietal and frontal cortex – '4 + 4 is 8. And 8 + 4 is 12.' Then they develop another way to outsource, automatizing the multiplication in a memory table in order to move on to more complex calculations.

An almost analogous process explains the virtuosos we looked at earlier. When chess masters solve complicated chess problems, what activates most differently is their visual cortex. We can summarize by saying that they don't think more, but rather they *see better*.* The same thing happens with great mathematicians who, when solving complicated theorems, activate their visual cortex. In other words, the virtuoso managed to recycle a cortex ancestrally devoted to identifying faces, eyes, movement, dots and colours in order to utilize it in a much more abstract realm.

---

* When the celebrated world chess champion José Raúl Capablanca was asked how many plays he calculated, he replied: 'Just one, the best one.'

Automatizing reading

The principle we can infer from the triangle experts explains what is perhaps the most decisive transformation in education: turning visual scribbles (letters) into the voices of words. Since reading is the universal window on to knowledge and culture, this gives it a special pertinence over the rest of the human skills.

Why do we begin to read at five years old and not at four or six? Is that better? Is it best to learn to read by breaking down each word into the letters that make it up or the other way around, by reading the whole word and associating it with a meaning? Given reading's pertinence, these aren't decisions that should be made from a stance of *it looks to me like*, but rather they should be built on a body of evidence that brings together the experience of years and years of practice and a knowledge of the cerebral mechanisms that support the development of reading.

As in other realms of learning, the expert reader also outsources. Those who read poorly not only read more slowly; what most holds them back is that their system of effort and concentration is focused on the reading and not on thinking about what the words mean. That is why dyslexia is often diagnosed by a deficit in reading comprehension. But it doesn't have anything to do with intelligence, rather simply with the fact that the effort is being put somewhere else. In order to be able to empathize with this, try to remember these words while you read the next paragraph: *tree, bicycle, mug, fan, peach, hat.*

Sometimes, when reading in a foreign language we've just started to learn, we come to realize that we've understood very little, because all of our attention was focused on translating. The same idea applies to all learning processes. When someone starts to study percussion, their focus is on the new rhythm they are learning. At some point that rhythm becomes internalized and automatic, and only then can they concentrate on the melody that floats above it, on the harmony that accompanies it or on other rhythms that are in a dialogue with that one simultaneously.

Do you remember the words now? And if you do, what was the last paragraph about? Successfully completing both tasks is difficult because each one occupies a limited system in the frontal and parietal cortex. Your attention chooses between juggling those six words so they don't vanish in your memory or following a text. Rarely can it focus on both.

The ecology of alphabets

Almost all children learn language very well. As a mature adult, I arrived in France without being able to speak more than the simplest words of French and it seemed strange to me that a small child I met who knew nothing of Kant's philosophy or calculus or the Beatles could speak French perfectly. It must surely have seemed strange to the same kid that a grown-up was incapable of something so simple as correctly pronouncing a word. This everyday example demonstrates how the human brain can display a mental virtuosity that has very little to do with other aspects we associate with culture and intelligence.

One of Chomsky's ideas is that we learn spoken language so effectively because it is built on a faculty for which the brain is prepared. As we have already seen, the brain is not a *tabula rasa*. It already has some built-in functions, and the problems that depend on them are more easily resolved.

In the same way that Chomsky argued that there are elements common to all spoken languages, there is also a common thread to all alphabets. Of course, the thousands of alphabets that exist, many already in disuse, are very different. But if we look at them all together, we immediately notice some regularities. The most striking is that their construction is based on just a few hand strokes. Hubel and Wiesel won their Nobel Prize for discovering precisely that each neuron of the primary visual cortex detects strokes in the small window it is sensitive to. The strokes are the basis of the entire visual system, the bricks of its form. And all alphabets are built with these bricks.

There are horizontal and vertical lines, angles, arches, slashes. And when you count the most frequent strokes in all alphabets there is extraordinary regularity: those strokes that are most common in nature are also the most common in alphabets. This isn't the product of a deliberate, rational design; alphabets just evolved to use a material that is quite similar to the visual material we are used to dealing with. Alphabets usurp elements for which our visual system is already fine-tuned. It's like starting with an advantage, since reading is close enough to what the visual system has already learned. If we try to teach with alphabets that have no relationship to what our visual system naturally recognizes, the experience of reading would be much, much more tedious. And, the other way around: when we see cases of reading difficulties, we can ease that process by making the material to be learned into something more digestible, more natural, more easily consumed – something for which the brain is prepared.

The morphology of the word

New readers pronounce a letter as if in slow motion. After many repetitions, this process becomes automatic; the ventral part of the visual system creates a new circuit able to recognize letters. This detector is built by recombining previously existing circuits that identify strokes. And in turn these become new bricks in the visual system that, like Lego pieces, are recombined to recognize syllables (of two or three successive letters). The cycle continues, with the syllables like new atoms of reading. At this point, a child reads the word 'father' in two cycles, one for each syllable. Later, when reading is firmly established, the word is read in just one sweep, whole, as if it were a single object. In other words, reading goes from being a serial process to a parallel one. At the end of the process of reading, readers form a brain function capable of extracting most words – except for extremely long and compound words – as a whole.

How do we know that adults read word by word? The first proof is in a reader's eyes, which move and stop on each word. Each one of these stops lasts about 300 milliseconds and then jumps abruptly and rapidly to the next word. In writing systems like English, that move from left to right, we focus very close to the first third of the word, since we mentally sweep from there to the right, towards the *future* of reading. This very precise process is, of course, implicit, automatic and unconscious.

The second proof is in measuring the time it takes to read a word. If we read it letter by letter, the time would be proportional to the length of the words. However, the time a reader takes to read a word made up of two, four or five letters is exactly the same. This is the great virtue of parallelization; it doesn't matter if there are one, ten, a hundred or a thousand nodes to which we must apply the operation. In reading, this parallelization has a limit in very long and compound words, such as sternocleidomastoid.* But within a range of two to seven letters reading time is almost identical. On the contrary, for someone who is just learning to read or for dyslexics, the reading time increases proportionally to the number of letters in a word.

We saw that the talent students have when starting their studies is not a good predictor of how good they will be after many years of learning. We now understand why.

In France, based on the finding that expert readers read word by word, one group concluded – erroneously – that the best way to teach children how to read was through *holistic reading* which, instead of starting by identifying the sounds of each letter, starts by reading entire words as a whole. This method quickly spread in popularity, probably because it has a good name. Who doesn't want their child to learn with the *holistic method*? But it was an unprecedented pedagogical disaster that led to many children having reading difficulties. And, with the argument sketched here, you can understand why the holistic method didn't work. Reading in a parallelized way is the final

---

* Luis Pescetti suggests that how natural a food is can be determined by counting the number of syllables in its name. Apple, peach, zucchini – all natural foods have fewer than five syllables.

phase that can only be reached by first constructing the intermediate functions.

The two brains of reading

Throughout this book we have focused on two different brain systems: the frontoparietal, which is versatile but slow and demands effort; and the ventral system, which is devoted to some specific functions that are carried out automatically and very quickly.

These two systems coexist, and their relevance varies over the course of the learning process. As seasoned readers, we primarily utilize the ventral system, although the parietal system is working residually, as becomes evident when we read complex handwriting or when the letters are not configured in their natural form, either vertically, from right to left, or separated by big spaces. In these cases, the circuits of the ventral cortex – which are not very flexible – cease to function. And then we read similarly to how a dyslexic person does.* In fact, it is harder to read a CAPTCHA** because it has irregularities that make the ventral system unable to recognize it. That is a way to dial back into the latent serial system of reading and find ourselves in the same situation we were long ago when we learned to read.

The temperature of the brain

When we learn, the brain changes. For example, the synapses – from the Greek for 'join together' – that connect different neurons can

---

* Try to read the following sentence backwards: 'A man, a plan, a canal, Panama.' It's a quite awkward way of saying the same thing, isn't it?
** CAPTCHA is an acronym (Completely Automated Public Turing test to tell Computers and Humans Apart) for an automated process that separates humans from machines. They are those drawn and camouflaged words that we have to type in to do many transactions on the Internet. Since computers cannot read these images, when we write them we are opening a lock only humans have the key to.

multiply their connections or vary the efficiency of an already established connection. All of this changes the neuronal networks. But the brain has other sources of plasticity; for example, the morphological properties or genetic expression of its neurons can change. And in some very specific cases, the number of brain cells can increase, although this is very rare. In general the adult brain learns without augmenting its neuronal mass.

Today the term 'plasticity' is used to refer to the brain's ability to transform. It is a popular metaphor but the term leads to an incorrect assumption: that the brain is moulded and stretched, gets crumpled up and smoothed out, like a muscle, although none of that actually happens.

What makes the brain more or less predisposed to change? With materials, the critical parameter that dictates their predisposition to change is temperature. Iron is rigid and not malleable, but when heated it can change shape and later be reconfigured into another form that it retains when cooled. What is the equivalent in the brain to temperature? First of all, as Hubel and Wiesel proved, there is the stage of development. A baby's brain does not have the same degree of malleability as an adult's. Yet, as we have seen, this is not an immutable variable. Is motivation the fundamental difference between a child and an adult?

Motivation promotes change for a simple reason that we have already discussed: a motivated person works harder. Marble is not exactly plastic but, if we go at it for hours with a chisel, it will eventually change shape. The notion of plasticity is relative to the effort we are willing to put in to make a change. But this doesn't yet bring us to the notion of temperature, of predisposition to change. What happens in the brain when we are motivated that predisposes it to change? Can we emulate this cerebral state in order to promote learning? The answer is in understanding which *chemical soup* of neurotransmitters promotes synaptic transformation and, therefore, cerebral change.

Before getting into the microscopic detail of brain chemistry, we should have a look at a more canonical way of learning: memory. Almost all of us remember the events of 9/11, the planes crashing at the North and South Towers of the World Trade Center. What's most surprising is that, fifteen years later, we not only remember those images of the burning towers but we also can recall with striking clarity where we were and who we were with when it happened. That deeply emotional moment makes everything around it, both the relevant – the attack – and the irrelevant, stick in the memory. This is why those who have been through a traumatic experience often have a very difficult time erasing that memory, which can be activated by fragments of the episode, the place where it happened, a similar smell, a person who was there, or any other detail. Memories are formed as episodes; in the moments our neuronal register is most sensitive, we vividly remember not only what activated that sensitivity, but also everything that formed around that episode.

This is an example of a more general principle. When we are emotionally aroused or when we receive a reward (monetary, sexual, emotional, chocolate) the brain is more predisposed to change. To understand how this happens, we have to switch tools and enter into the microscopic world. And that voyage will take us to California, to the laboratory of a neurobiologist, Michael Merzenich.

In his experiment, monkeys had to identify the higher pitched of two tones, as when we tune an instrument. As the two tones became more similar, they began to perceive them as identical even though they weren't. This allowed him to investigate the limits of the resolution of the auditory system. Like any other virtue, this can also be trained.

The auditory cortex, just like the visual one, is organized into a pattern, a cluster of neurons grouped into columns. Each column is specialized to detect a particular frequency. So, in parallel, the auditory cortex analyses the frequency structure (the notes) of a sound.

In the map of the auditory cortex, each frequency has a dedicated region. Merzenich already knew that if a monkey is actively trained to recognize tones of a particular frequency, something quite extraordinary happens: the column representing that frequency expands, like a country that grows by invading its neighbouring territories. The question that concerns us here is the following: what allows that change to happen? Merzenich observed that the mere repetition of a tone wasn't enough to transform the cortex. Yet if that tone occurs at the same time as a pulse of activity in the ventral tegmental area, a region deep in the brain that produces dopamine, then the cortex reorganizes itself. It all comes together. In order for a cortical circuit to reorganize, there needs to be a stimulus occurring in that window of time which releases dopamine (or other similar neurotransmitters). In order for us to learn, we need motivation and effort. It isn't magic or dogma. We now know that this produces dopamine, which lessens the brain's resistance to change.

We can think of dopamine as the water that makes clay more pliable, and the sensory stimulus as the tool that marks a groove in that damp clay. Neither can transform the material on its own. Working dry clay is a waste of time. Wetting it if you aren't going to sculpt it is as well. This is the basis of the learning programme that we began discussing with Galton's idea: the brain learns when it is exposed to stimuli that transform it. It is a slow and repetitive task to establish those grooves of new circuits that automatize a process. And the transformation requires, in addition to effort and training, a cerebral cortex that is in a state sensitive to change.

To sum up, we looked at Galton's error in order to understand how learning is forged: the ceiling is not as genetically established as is commonly assumed and the path is also social and cultural. We also saw that virtuosos carry out their expert tasks in a qualitatively different way, not only by improving the original procedure. And that to persevere in learning we have to work with motivation and effort, outside our comfort zone and the OK threshold. What we recognize as a performance ceiling is usually not. It is an equilibrium point.

In short, it is never too late to learn. If something changes in adulthood it is that our motivation gets stuck in what we've already learned and not swept up in the whirlwind of discovering and learning. Recovering that enthusiasm, that patience, that motivation and that conviction seems to be the natural starting point for those who truly want to learn.

CHAPTER SIX

Educated brains

How can we use what we have learned about the
brain and human thought to improve education?

Every day, more than two billion children around the world go to
school, in what is perhaps the largest collective experiment in the
history of humanity. There they learn to read, forge their closest
friendships, and build themselves as social beings. And at school, in
a highly intense learning process, the brain is developed and transformed. However, neuroscience has crassly ignored this close link,
remaining distanced from classrooms for years. Perhaps now is the
prime time to establish a bridge between neuroscience and education.

The philosopher and educator John Bruer warned that this bridge
connects distant worlds; what neuroscience considers relevant is not
necessarily or usually pertinent to education. For example, understanding that a region of the parietal cortex is key to the numerical
process can be important for neuroscience but doesn't help a teacher
to teach maths better.

In this effort, we must remain, more than ever, sceptical about the
use of vague and imprecise scientific terms. I was once in a conference
in which a supposed expert in neuroscience argued – as so many do
today – that we should use the right hemisphere more. I raised my
hand (the left one to be compliant . . .) and mentioned that even if I

agreed that using the right hemisphere was useful, I simply did not know how to do it. Should I turn my head to the right to increase blood flow to the right hemisphere? His 'neuro-expert' response was that I should focus on drawing, colouring books and creative arts and forget about language. And then my question was, why didn't he just say that straight out? I knew the answer. He was employing an artful yet useless metaphor. Referring to the brain and the hemispheres only served to appropriate the prestige of a scientific field for marketing purposes.

There is a long history of translating basic knowledge into applied science. One perspective maintains that science should produce a body of knowledge with the hope that some of it will eventually be useful for society's needs. An alternative approach, coined by Donald Stokes as *Pasteur's quadrant*, consists in finding a niche where basic and applied research meet.

In Stokes's taxonomy, scientific knowledge is classified according to whether it searches for fundamental understanding or has immediate use for society. The atomic model by Niels Bohr, for example, is a case in which science chases after pure knowledge. On the other hand, Thomas Edison's light bulb is an example that takes usage into account. Pasteur's research on vaccination, according to Stokes, deals with both dimensions; in addition to resolving fundamental principles of microbiology, it gave a concrete solution to one of the most urgent medical problems of the age.

In this chapter we will try to navigate the waters of neuroscience, cognitive science and education along Pasteur's quadrant, exploring fundamental aspects of brain function in the hope of contributing to the quality and efficacy of educational practice.

The sound of the letters

When we learn to read we discover that the shapes p, *p*, 𝒫, *p*, ρ and P are the same letter. We understand that the precise combination of a line segment and a curve, of the '| + ⊃', makes up the P. The curve can be smaller, the line can be tilted and the curve can slightly cross it, but we know that these forms, which are never identical, represent the same letter. This is the visual part of reading, whose process we have already looked at. But there is another, more complicated action, which entails learning to pronounce it. Understanding that this visual object 'p' corresponds to an auditory object, the phoneme /p/.

Consonants are difficult to pronounce because we never hear them isolated; they are always accompanied by a vowel. That's why the consonant 'p' is called 'pee'. Naming it without the 'ee' that follows feels strange. Additionally, some consonants require complex morphologies of the vocal apparatus like the explosive union of the lips to produce the /p/ or the palate juncture needed to produce the /j/. Syllables, especially when they are comprised of a consonant and a vowel, like 'pa', are much easier to pronounce.*

In Spanish or Italian there is a precise correspondence between phonemes and letters, which makes decoding them fairly transparent. But in English and in French that doesn't happen, and those who are learning to read have to decipher a less straightforward code that forces them to scan a few letters before they can know how to pronounce them.

The importance of the expressive component of reading is usually underestimated, in part, perhaps, because we can read in silence. But even if we are reading *in a whisper*, we advance more slowly when the words are harder to pronounce. Which is to say, we internally pronounce the text we are reading even when we produce no sound.

Therefore, those who are learning to read are also discovering how to speak and how to listen. When pronouncing the word 'Paris' we

---

* In English, syllables usually have a complex structure. In Spanish and Italian, on the other hand, the simple consonant-vowel structure is frequent, and it is even more common in Japanese. That is why the Japanese have such difficulty pronouncing, when they appear in other languages, syllables ending in a consonant, saying 'aiscrimu' and 'beisoboru' for *ice cream* and *baseball*.

produce a continuous stream of sound.* Asking someone who doesn't know how to read to divide the word into /p/ /a/ /r/ /i/ /s/ is like trying to separate a ball of used, mixed Play-Doh into its pure original colours. Impossible. The syllables, and not the phonemes, are the natural building blocks of the sounds of words. As such, without having learned to read it is very hard to answer the question of what happens if we take the 'P' off the word 'Paris'. This ability to break up the sound of a word into the phonemes that comprise it is called phonological awareness and is not innate but rather acquired along with reading.

Reading trains phonological awareness because in order to recognize a phoneme as a building block of speech it has to have a label, a name that distinguishes it and turns it into an object within that stream of sound. These labels are precisely what make up the letters that a phoneme represents. Therefore, an essential part of reading is discovering phonemes. In fact, most reading difficulties are not visual but auditory and phonological. Ignoring the phonological aspect of reading is one of the most frequent misperceptions in teaching.

Word-tied

Dyslexia is perhaps the most paradigmatic example of how neuroscience can be useful to education. First of all, research on the brain has helped us to understand that dyslexia has little to do with motivation and intelligence, but rather is the result of a specific difficulty in the cerebral regions that connect vision with phonology. The fact that dyslexia has a biological component doesn't mean that it cannot be improved or reversed. It is not a stigma. Quite the opposite: it allows us to understand an inherent difficulty that a child may have when learning to read.

Another typical error is thinking that the problem of dyslexia is in the eyes, when the greatest difficulty is usually in recognizing and

---

* And champagne.

pronouncing the phonemes; in other words, in the world of sounds. This discovery opens the door to simple and effective activities for improving dyslexia. The way to help dyslexic children is often not by working with their vision but rather by helping them to develop phonological awareness. Having them listen to and pick up on the differences between 'paris, aris, paris, aris . . .' for example. In fact, this game of deleting a phoneme from a word is an excellent reading exercise: 'Starling, staring, string, sting, sing-sin-in-I.'

Neuroscience can also help to recognize dyslexia before it is too late. Sometimes it only becomes obvious that a child is having a specific difficulty with reading after valuable months or years of his or her educational experience have already passed. With dyslexia, as in many other realms of medicine, early detection can radically change the prognosis. But the same medical analogy works to warn of the obvious, that this is a very delicate subject which requires special care and prudence. There is a clear advantage to early diagnosis, but the risk of stigmatization and self-fulfilling prophecy is also evident.

This decision becomes particularly hard because dyslexia cannot be predicted with certainty; we can only infer a predisposition to it. Let's look for a moment at a more concise example, congenital deafness. Without mediating science, deafness is diagnosed later because during the first few months of a baby's life the fact that they don't respond to sounds goes unnoticed. With early detection, however, the baby's parents can start to use a gestural, symbolic language and essentially a deaf baby will grow up better able to communicate. That child's world will be less wide and strange. In fact, medical practice has already radically changed to recognize this awareness of the importance of early diagnosis. Soon after birth, babies are given an acoustic test that indicates whether they have an auditory dysfunction. With an early diagnosis of possible deafness, parents can be attentive to those aspects and improve their children's social development. Something similar happens with dyslexia: the cerebral response to phonemes – at one year old – is indicative of the difficulties babies might encounter almost four years later, when they begin learning to read.

The subject is so sensitive and delicate that it is tempting simply to turn a blind eye to it. But ignoring this information is also a way of deciding. Decisions made by default – by not doing anything – might feel easier to make but do not side-step the need to take responsibility. One thing is for sure, a near future in which we will be able to estimate the likelihood that a child will develop dyslexia is imminent. What we must decide – at all levels of society, from parents, to teachers and head teachers, to policy makers – is how to act on this information. And this of course is a decision that goes beyond the scope of science.

My opinion is that information about the likelihood of dyslexia can be used carefully and respectfully, without stigmatizing children. It is good for parents and educators to know if a child has a significant probability of having difficulties in reading. This will allow them to give the child the opportunity to do some phonological exercises (which are completely innocuous and even entertaining) that might help in overcoming that initial disadvantage, in order to learn how to read, so that they have better prospects when starting the first year of school, with the same possibilities as the rest of their classmates.

To sum up:

(1) Phonological awareness, which has to do with sound and not sight, is a fundamental building block of reading.
(2) There is much initial variation in that ability – before starting to read, many children already have a configuration of their auditory system that naturally separates phonemes, while others have them more mixed up. Children who have low resolution in their phonological systems show a predisposition for dyslexia.
(3) With harmless and fun activities, like simple word games, the phonological awareness system can be stimulated before reading begins, at two or three years old, so that those children don't start to learn to read facing a disadvantage.

The study of reading development is one of the most prominent cases of the way in which investigation of the human brain can be useful to educational practice. It is at the core of this book's intention to explore how this reflective exercise on the part of science can help us to understand ourselves and communicate better.

What we have to unlearn

Socrates questioned what common sense suggests, that learning consists of acquiring new knowledge. Instead, he proposed that it involved reorganizing and recalling knowledge we already have. I now put forth an even more radical hypothesis of learning understood as a process of editing, as opposed to writing. Sometimes, learning is losing knowledge. Learning is also forgetting. Erasing things that take up space uselessly and others that, even worse, are a hindrance to effective thought.

Young children usually write some backwards letters. Sometimes they even write a word or an entire sentence as if in a mirror. Compared to other 'mistakes' that children make when learning, this one is often overlooked, like some sort of endearing temporary clumsiness. But actually it is an extraordinary feat. First of all, because the children were never taught to write backwards. They learned it on their own. Secondly, because mirror writing is very difficult. In fact, just try to write an entire sentence backwards, the way kids do naturally.

Why does the development of writing have this peculiar trajectory? What does this teach us about how our brain works? The visual system converts light and shadow into objects. But since objects turn and rotate, the visual system is not very interested in their particular orientation. A coffee mug is the same turned backwards. Almost the only exceptions to this rule are certain cultural inventions: letters. The mirror reflection of 'p' is no longer a 'p' but a 'q'. And if we reflect it upside down it becomes a 'd' and then left to right again it turns into a 'b'. Four mirrors, four different letters. Alphabets inherit the same fragments and segments of the visual world, but their symmetry is an exception. The reflection of a letter is not the same letter. That is atypical and unnatural for our visual system.

In fact, we have a very poor memory for the particular configurations of objects. For example, almost everyone remembers that the Statue of Liberty is in New York, that it is somewhat greenish, that it has a crown and one hand raised with a torch. But is the torch hand the left or the right? Most people can't remember which it is, and those who think they do are often wrong. And which way is the *Mona Lisa*'s gaze directed?

It makes sense that we forget those particular details, since our visual system has to actively ignore these differences in order to identify that all the rotations, reflections and shifts of an object are still the same object.* The human visual system developed a function that distinguishes us from *Funes the Memorious* and makes us understand that a dog seen in profile and a dog seen head on are the same dog. This highly effective circuit is ancestral. It worked in the brain long before schools and alphabets existed. It was later in the history of humanity that alphabets appeared, imposing a cultural convention that goes against the grain of our visual system's natural functioning. According to this convention, 'p' and 'q' are two different things.

Those who are learning to read still function with a default setting in their visual systems, in which the 'p' is equal to the 'q'. Therefore they are naturally confused both in reading and in writing. And part of the process of learning implies uprooting a predisposition, eradicating a vice. We have already seen that the brain is not a *tabula rasa*

---

* This was clearly and concisely expressed by Jorge Luis Borges in *Funes the Memorious*. 'Not only did he struggle to understand that the generic symbol dog covered so many disparate individuals of diverse sizes and diverse shapes; it bothered him that the dog from three fourteen (seen in profile) had the same name as the dog from three fifteen (seen head on). His own face in a mirror, his own hands, surprised him every time [. . .] I nevertheless suspect that he was not very capable of thought. Thinking is forgetting differences, it is generalizing, abstracting.'

where new knowledge is written. And as we just saw in the case of reading, some spontaneous forms of functioning can result in idiosyncratic difficulties in learning.

The framework of thought

From the day we are born the brain already forms sophisticated conceptual constructions, like the notion of numerosity, and even morality. We root our reconstruction of reality in those conceptual boxes. When we listen to a story, we don't record it word by word but rather we reconstruct it in the language of our own thoughts. That is why people emerge from the same cinema with different stories. We are the scriptwriters, directors and editors of the plot of our own reality.

This is highly pertinent in the educational environment. The same thing that happens with a film occurs with a class; each student reconstructs it in their own language. Our learning process is a sort of convergence point between what is presented to us and our predisposition for assimilating it. The brain is not a blank page on which things are written, but rather a rough surface on which some shapes fit well and others don't. That is a better metaphor of learning. A problem of congruity, of matching.

One of the most exquisite examples is the representation of the world itself. The Greek cognitive psychologist Stella Vosniadou studied thousands and thousands of drawings in detail to reveal how children's representations of the world change. At some point in their educational history, children are presented with an absurd idea: the world is round. The idea is ridiculous, of course, because all factual evidence accumulated over the course of their lives points to the opposite.*

---

* John Lennon knew something about this: 'Because the world is round it turns me on.'

In order to understand that the world is round one must unlearn something very natural based on sensory experience: the world is clearly flat. And when we understand that the world is round, other problems begin. Why don't people in China, on the other side of the world, fall off? Here gravity starts to do its job, keeping everybody stuck to the earth. But this in turn brings new problems; why doesn't the world fall if it is just floating in space?

The conceptual revolutions we experience throughout our life emulate, to a certain extent, the development of culture in history. The children who are shocked when they hear that the world is round are replicating the conceptual struggle of Queen Isabella when Columbus suggested his voyage to her.* So the problem of the earth floating in the middle of nothing is resolved in young infancy as it was so many times in the long history of human culture, resorting to giant turtles or elephants that hold it up. Beyond the fable, what is interesting is how each individual has to find solutions to resolve a construction of reality according to the conceptual framework in which they find themselves. An expert physicist can understand that the world is spinning, that it has inertia, that in reality it is in an orbital motion, but an eight-year-old cannot solve the dilemma of why the world doesn't fall with the arguments in a child's arsenal.

For classroom teachers, parents or friends, it is very useful to know that those who are learning assimilate information in a very different conceptual framework from their own. Pedagogy becomes much more effective when that is understood. It is not about just speaking more simply but rather about translating what you know into another

---

* Although most likely that conversation never happened. It is a myth invented in modernity that all those in medieval times believed that the earth was flat. Aristotle had already proven that the earth is spherical, and everyone accepted it (Eratosthenes even measured its size). It was something that any medieval person who was averagely educated knew. It is an incredibly widespread modern invention that Columbus was the bold one who wanted to try to prove that. This story is told in *Inventing the Flat Earth: Columbus and Modern Historians* by J. Russell (New York, Praeger, 1997).

language, another way of thinking. That is why, paradoxically, sometimes teaching improves when the teacher is another student who shares the same conceptual framework. At other times, the best translators are the students themselves.

The mathematicians Fernando Chorny, Pablo Coll and Laura Pezzatti and I did an extremely simple test, but which may have important consequences for educational practice. We put a mathematical problem to hundreds of students who were preparing for an exam in an entry-level course after dividing them into two groups. The first group was simply asked to solve the problem, just as with any other test. The second group was asked first to rewrite the formulation of the question in their own words and only then to solve the problem.

From one perspective, the extra task for the second group was a distraction that meant they had less time and concentration. But from the perspective that we sketched out here, it pushed them to do something key to learning: translating that formulation into their own language before solving it.* The change was spectacular; the performance of those who rewrote the problem

---

* This was the word problem. You can rewrite it and you'll see how much easier it is to solve. 'The floors of a building are numbered from 0 to 25. The building's elevator has only two buttons, one yellow and one green. When the yellow button is pressed it goes up 9 floors, and when the green button is pressed, it goes down 7 floors. If the yellow button is pushed when there are not enough floors above, the elevator will not move, and the same thing happens when the green button is pushed and there aren't enough floors below. Write a sequence of buttons that will allow a person to go up from floor 0 to 11 in the elevator.' And this is my translation, written almost in code, which allowed me to solve it much more easily without uselessly saturating my memory buffer:
Elevator: up 9 or down 7.
Building: 25 floors.
You cannot go past the ground level or the roof.
How can you get from 0 to 11?

improved almost 100 per cent over those who directly solved the problem as we had put it forth.

Parallelawhat?

Now we will look at the world of geometry from the perspective of a child in order to discover that the process of rewriting concepts in one's own language goes far beyond the world of words. In fact, it is enough to read the definition of parallelism to understand that geometry doesn't get along well with words: 'Equidistant from another line or plane, so that no matter how long they extend, they cannot intersect.' The definition is filled with abstract terms: line, plane, equidistant (often the concept of infinity is used to define it as well). The word itself – 'parallel' – is complex to pronounce. Who would take to something like that? Yet, when we see two lines that are not parallel among several that are, they immediately pop out at us. Our visual system establishes intuitions that allow us to recognize geometric concepts long before they are put into words.

Three-year-olds can already distinguish two non-parallel lines among many parallel ones. Perhaps they can't explain the concept, much less name it, but they understand that there is something that makes those lines different. The same thing happens with many other geometrical concepts, the right angle, closed or open figures, the number of sides on a figure, symmetry, etc.

There are two natural ways to investigate universal aspects that are not established by education. One is by observing children before they have been overly affected by culture and the other is by travelling to places where education is very different, as a sort of anthropologist of thought.

One of the most studied cultures for investigating mathematical thought is that of the Munduruku tribe, deep in the Brazilian Amazon. The Munduruku have a very rich, ancient culture, with very different mathematical ideas from those we inherited from the Greeks and Arabs. For example, they don't have words for most numbers. There is a composite word to refer to one (*pug ma*), another for two (*xepxep*), another for three (*ebapug*), another for four (*ebadipdip*), and that's it. Then they have words that represent approximate amounts, like *pug pogbi* (a fistful), *adesu* (some) and *ade ma* (quite a few). In other words, they have a mathematical language that is more approximate than exact. Their language can distinguish between *many* and *few* but not determine nine minus two is seven, which is inexpressible. Seven, thirty and fifteen do not exist in the Munduruku language.

Nor is their language rich in abstract geometrical terms. Does that mean that geometric intuitions in the Munduruku communities are very different from those in Boston? The answer is no. The psychologist Elizabeth Spelke discovered that when geometric problems are expressed visually and without using language, Munduruku children and kids from Boston solved them with very similar results. What's more, the things that are simple for a kid from Boston – like recognizing right angles among other angles – are also easy for a Munduruku. Harder things – like recognizing symmetrical elements among non-symmetrical ones – are difficult for both groups of children.

Mathematical intuitions cut across all cultures and are expressed from infancy. Mathematics is built on intuitions about what we see: the big, the small, the distant, the curved, the straight; and about space and movement. In almost all cultures, numbers are expressed in a line. Adding is moving along that line (typically towards the right) and subtracting is doing the same thing in the other direction. Many of these intuitions are innate and develop spontaneously, without the need for any formal instruction. Later, of course, formal education is added on top of that body of already formed intuitions.

When comparing adults in Boston with Mundurukus, the former solved geometric problems much more effectively. This is almost stating the obvious, merely corroborating the fact that, if someone spends years studying a trade, they get better at it. But what's most interesting and revealing is that while education improves our ability to solve all the problems, there is still a hierarchy of difficulty. The problems that are most difficult for us as adults are the ones that were impossible when we were children.

To sum up, when people discover something, they analyse it according to their own conceptual framework, which is built from very early (maybe even innate) intuitions. Through time and learning we go through conceptual revolutions that change the way in which we organize concepts and represent the world. But old intuitive conceptions persist. And we can trace that childish way of solving problems through adulthood, even in proficient experts considered to be great thinkers in their field. Problems that are not very intuitive remain tedious and hard to solve throughout our educational formation. Understanding how this body of intuitions works within the human mind is one natural route to improve the way we teach our children.

Gestures and words

Earlier, I described learning as a process that transfers reasoning to the visual cortex of the brain in order to make it parallel, fast and efficient. Now we will look at the inverse process by which we acquire symbols that can describe innate visual intuitions.

Liz Spelke, Cecilia Calero and I studied how geometric intuitions turn into rules and words. Our theory was that the acquisition of knowledge has two stages. The first is a hunch; the body knows the response but cannot express it in words. Only in a second stage do the reasons become explicit as rules that can be described to ourselves and others. We also had another theory, conceived in the desert of Atacama, where Susan Goldin-Meadow, one of the great researchers of human cognitive development, told us about an extraordinary discovery she made after re-examining an old exercise done by Jean Piaget.

In the Swiss psychologist's experiment, children were shown two rows of stones and had to choose which one had more. The trick was that while both rows had the same number of stones, in one of them they were more spaced out. The six-year-olds, driven by a ubiquitous intuition in our thinking, confused length with quantity and systematically chose the longer row.

Susan made a subtle but very important discovery about this classic experiment. While all the children answered that there were more stones in the longer row, there were remarkable differences in how they gestured their response. Some extended their arms to show that one row was much longer than the other. Other children moved their hands to establish a correspondence between the stones in each row. Those children, who were counting with their hands, had in fact discovered the essence of the problem. They weren't able to express that knowledge with words, but their body language did. For that second group of children, the Socratic dialogue would work. The teacher only has to give them a little push to help them express the knowledge they already have. This finding is not a mere intellectual curiosity; when educators apply this information, their teaching becomes much more effective.

By this careful observation, Susan discovered that gestures and words tell different stories. We then decided to explore how the children expressed their geometric knowledge along three different channels: their choices, their explanations and their gestures.

In our experiment, the children were asked to choose the odd man out among six cards, the only one that didn't share a geometric property with the others. For example, five of the cards had two parallel lines drawn on them and the other had two oblique lines in the shape of a V. More than half of the children under four years of age chose the only card that showed non-parallel lines. The others chose wrongly, but not randomly.

Some chose the card that had the most space between the two lines. Or the one in which the lines were the longest. They were focusing on an irrelevant aspect of the problem. Most of those children explained their choice in a consistent way, using words that referred to size. Their actions were coherent with their words. However, their hands told a completely different story. They moved them to form a wedge shape and then in parallel. Which is to say, their hands clearly expressed that they had discovered the pertinent geometric rule. Let's just say that if it were an exam, their spoken answer would have failed them, but if they were scored on their hands they'd have passed.

We do not know yet the brain mechanisms that explain why information about geometry may be expressed through gestures or choice but not through language. Or what exactly happens in the brain in the moment in which children can have a more consistent grasp of these geometric intuitions and are able to express them in words.

But the experiments in which knowledge is measured through words, actions and gestures help us understand how we learn to forge concepts. Some concepts, like shape, form part of a core set of intuitions that are accessible to implicit knowledge and only later in development can be conveyed explicitly. Younger children can easily identify an odd shape even when they cannot express (to others and probably also to themselves) the geometric reasoning that justifies these choices.

The development of other geometric concepts, such as angles, follows a different path. They are first expressed through gestures at a time in which children cannot use this information for solving specific problems or for describing these concepts in words.

Why different concepts build up differently may be due to our innate biological predisposition, but most certainly, as well, they are due to how we relate to geometry in schools and homes. Most children grow up playing frequently with shapes, but have very little practical experience with angles, a dimension that can be much more naturally expressed by gestures. Above and beyond this, the more general point is that there are different precursors that serve to consolidate explicit knowledge.

Cecilia's study showed how rudimentary children are when they have to express geometrical concepts with words. And in fact it's not just children that this is true for. The Menon dialogue, which I described at the beginning of Chapter 5, shows that it's also the case for adults. Developing notions of geometry is different from many other concepts such as number or theory of mind, because geometrical concepts aren't composed in the same way that numerical and mental state concepts are. This is why it may be so hard for children and adults to express them verbally or learn them from others' verbal expressions.

And here is where the real pertinence of these results for educational practice becomes evident. First, they suggest that geometry (and many other concepts) may not be taught well using words. This might be the essence of the failure of the Menon dialogue. Second, they also tell a teacher that language may not be a good vehicle to inquire about students' knowledge of these matters.

The body is a consortium of expressions. Our words represent only a small fragment of what we know. And they are incredibly effective in conveying certain concepts and quite clumsy in expressing others. This may seem trivial in other domains. Imagine a football player being examined through a verbal description of how to take a free kick. As absurd as this may seem, it may be, to some extent, what we do with millions of children when we ask them to explain in words what they know about geometry.

Good, bad, yes, no, OK

Luis Pescetti, an Argentinian novelist, musician and actor, wrote a song in which a parent asks a teenage son a long series of questions. They all have the same responses: yes and no. This, of course, doesn't mean that the son has no replies to the questions; just that he doesn't want to answer. The song touches on an important lesson for developmental science: the best way to discover a teenager's or a child's inner thoughts is not through direct questioning, not in real life and not in the realm of scientific experimentation.

By exploring various procedures for investigating what children know, we found that the best way was not to ask anything but just to let them speak. This reveals an important principle of social beings: nothing has meaning in and of itself, but, rather, meaning is acquired when someone can share it. The need to share and communicate is a very natural predisposition.

What began as a technical resource for investigating explicit knowledge became something much more interesting, since we discovered that children have a sort of *teaching instinct*. They are natural teachers. A child with any sort of knowledge has a very strong propensity to share it.

The teaching instinct

Antonio Battro studied with Piaget in Geneva in 1967. Over time he became the standard-bearer of technological transformation in the classrooms of Nicaragua, Uruguay, Peru and Ethiopia. Just as we were exploring children's innate desire to share their knowledge, Antonio came to our laboratory in Buenos Aires with an idea that was to transform our work, protesting that it was absurd that all neuroscience was dedicated to studying how the brain learns while completely ignoring how it teaches. And he argued that this was particularly strange because the ability to teach is one of the things that distinguishes us as a species, that makes us human. It is the seed of all culture.

We share the capacity to learn with all other animals, including the *Caenorhabditis elegans*, a worm less than a millimetre long, and the *Aplysia* sea slug, with which the Nobel laureate Eric Kandel discovered the molecular and cellular mechanics of memory. But we have something distinctive and particular that takes this ability and both communicates and propagates knowledge; those who have learned something have the capacity to transmit it. It is not a passive process of assimilating knowledge. Culture travels like a highly contagious virus.

Our hypothesis was that this voracity to share knowledge is an innate compulsion, like drinking, eating or seeking pleasure. To be more precise, it is a programme that develops naturally, with no need to be taught or explicitly trained. We all teach, even when no one has ever taught us how. Just as Noam Chomsky suggested we have an instinct for language, my colleague and friend Sidney Strauss and I emulated his idea and proposed that we all have a teaching instinct. The brain is predisposed to spread and share knowledge. This hypothesis is built upon two premises.

(1) Prototeachers

Long before learning to speak, children communicate. They cry, they request, they demand. But do they communicate information with the sole objective of remedying a gap in knowledge? Do they teach before starting to talk?

Ulf Liszkowski and Michael Tomasello came up with an ingenious game to answer these questions. An actor let an object fall off a table in full view of a one-year-old child. The scene was composed in such a way that the children saw where it fell but the actor didn't. Later, the actor diligently and fruitlessly searched for the object. The little ones spontaneously acted as if they recognized this gap in knowledge and wished to remedy it. And they did so with the only resource available to them (since they could not yet speak), which was by pointing to the location of the object. This could be merely automatism. But the most revealing element of this experiment was that, if it was made clear in the staging that the actor knew where the object had fallen, then the one-year-olds would no longer point to it.

This is almost pedagogy, in that:

(1) The infant does not gain anything (evident) from it.

(2) It denotes a clear and precise perception of a gap in knowledge.

(3) It isn't automatism but rather expresses a specific action in order to transmit knowledge to someone who does not have it.

In some sense, the one-year-olds have an economic perspective of knowledge; that is to say, the effort of transmitting it is only worthwhile when it is useful to the other person.

What their action lacks to make it fully teaching is for the transmission of knowledge to empower the student to continue on their own. In this case, the baby shows the actor where the object has fallen, but – *ungenerously* – doesn't show them how to find it when it falls again.

Before learning to speak, children can also proactively intervene by warning an actor when they anticipate their making a mistake. Which is to say, they try to close the communication gap even when dealing with actions they presuppose will happen but which have yet to occur. This ability to foresee others' actions and act accordingly is at the core of teaching and is expressed even before a baby starts to talk and walk.

(2) Teaching, naturally

No one taught us to teach as children. We obviously didn't go to teachers' college or pedagogical workshops. But if we indeed possess an innate teaching instinct, we should teach naturally and effectively. At least as children, before that instinct atrophies. Here we see a problem: the teaching quality depends on how much the teacher knows about the topic. To know whether children communicate effectively, independent of their specific knowledge about the subject, we have to observe their gestures, not their words. Here, what is not said is more important than what is.

There are universal aspects to human communication. Beyond words, semantics and content, one of the virtues of effective speeches

– like those of the great leaders in history – is that they work on an ostensive level. Ostensive communication is a concept that has been visited and revisited by philologists and semiologists like Ludwig Wittgenstein and Umberto Eco. It refers to the ability to use gestures to amplify the speech and use as few words as possible. It uses an implicit key that is shared between the speaker and the interlocutor. If we lift one hand with a salt shaker in it and ask someone: 'Want some?', there is no need to be explicit about what we are offering them. It's the salt. This is a precise dance of gestures and words that happens in a fraction of a second without us even knowing that we are dancing. A robot versed in language would have asked: 'Excuse me, what is it you are asking me whether I want some?'

The key to this method of communication is pointing. When we say: 'That one', and point, others understand what those words and that hand are indicating. It is a highly efficient way of communicating. Monkeys, who are able to do a countless number of sophisticated things, don't understand this code that is so simple for us. It is a way of relating to each other that defines us, that makes us who we are.

By adorning our speech with prosody, gestures and signs, 'ostensive communication' also serves to label and parse out relevant moments of discourse. With this, the emitter ensures that the listener does not get distracted during the essential part of a message, which would result in a major communication failure.

Ostensive keys are easily recognizable. One is looking into the other person's eyes and directing one's body towards them. Aiming one's gaze or body at the listener functions as a magnet for their attention. Other ostensive cues are using the receiver's name, lifting our eyebrows or changing our tone of voice. These all make up a system of gestures, which we recognize as natural but that were never taught to us, and that determine the efficiency with which a message is communicated. Perhaps the most spectacular demonstration of how gestures come naturally without needing to be taught is that they are used by the congenitally blind even when in many cases they have never perceived them through other sensory modalities. We can think of it as a channel of communication. The transmission of the message is effective if we tune that channel in well, and it becomes static-y, confused or ineffective if we don't find the exact frequency of this natural channel of human communication.

Two Hungarian researchers, Gergely Csibra and György Gergely,* discovered that the ostensive channel of human communication is effective from the very day we are born. Newborns not only learn *more* when we communicate while looking at them, changing our tone of voice, calling them by their name or pointing at relevant objects. They also learn in a completely different way.

When a message is communicated ostensively, the receiver understands that what they've learned goes beyond the particular case that is shown. When we tell babies without ostension that an object is a pencil, they understand it as a description of a particular object. Yet when we say the same thing with ostensive cues, they grasp that this explanation refers to a whole class of things that this object in particular belongs to.

When a message is communicated ostensively, receivers also assume that what they've been shown is complete, that the class is over. In an experiment illustrating this, a teacher shows children one of the many uses of a toy. In one case, this demonstration is carried out ostensively, with a gesture to finish that clearly indicates that the show is over. In the other case, after the demonstration, the teacher abruptly leaves the room.

In both cases the children were taught exactly the same thing, but their responses are very different. In the first case, the children do not explore other uses of the toy, denoting that they

---

* It's fabulous that these two extraordinary Hungarians, who revealed the mysteries of human communication, are linked, in that the last name of one is the first name of the other. Now we need a record by the singers Luis Miguel and Miguel Mateos, and the now impossible trio of Boy George, George Michael and Michael Jackson.

understand that the lesson was complete. In the second case, they spontaneously explore the toy's other functions, showing that they understand that they were explained only some of its uses.

At six years old, children make highly precise evaluations, based on ostensive cues, of the quality of the information they receive from a teacher. When they have reasons to doubt a teacher's reliability – for example, because of lack of ostension – they investigate beyond what they've been taught. So learning not only depends on the content of the message but also on the reliability of the person communicating it. This also reveals a paradox in education: good teachers transmit completeness and with this they inhibit their students' further exploration.

Gergely and Csibra gave this implicit code for sharing and assimilating information the designation 'natural pedagogy'. In other words, ostension is a natural and innate way of comprehending what is pertinent and relevant. This makes it possible to discover rules in a world of information as vast and as ambiguous as ours. Herein lies something essential to human intuition and comprehension, something that is very difficult to emulate and that explains the seemingly clumsy learning abilities of the automatons we design.

This survey of the fundamentals of human communication will now allow us to tackle the question we sketched out earlier. In order to know whether children are objectively good teachers, all we need to ask ourselves is whether they are ostensive, whether when communicating something important they lift their eyebrows, use the receiver's name, and direct their body towards them, using the entire arsenal of ostensive cues that will make the receiver pay attention and feel that the information transmitted is complete and reliable. And this is independent of whether or not what's transmitted is correct, which depends on how much they know about the subject and not how well they teach. It is a precise and implicit way of asking if they have well-formed intuitions about the effective channels of human communication. The path was left clear but we still needed to walk down it. Which is what we set forth to do with Cecilia Calero.

Our project involved a relatively simple arrangement, whose originality consisted in putting children in the place of the teachers. A child learned something, like a game, a mathematical concept, a universe with its own rules, or fragments of a new language. Then another person, who lacked that knowledge, would arrive on the scene. And from there we began to observe. In some cases we studied the children's propensity to teach the new arrivals. In others, the newcomers would ask for help, and we studied what, how and how much the child taught them.

We discovered that the children naturally taught with enthusiasm and loquacity. They smiled and enjoyed teaching. In the hundreds of activities that Cecilia did, there were many times when the children wanted to interrupt – and did – while they were learning. But there was not a single child who didn't want to teach.

During the class that the child gave the newcomer, there were moments of varying pertinence. Some were irrelevant to the exchange. For example, there was the boy who talked about his sister, that it was raining, or hot – weather is maybe the only topic we are comfortable talking about with any stranger, at any time, anywhere in the world. And other times one would transmit content relevant to the game they wanted to teach, such as its logic or strategy. And right at that moment the child teacher began using a barrage of ostensive cues. That display of gestures denoted that the child knew how to teach in order to gain the attention of the learner's more sensitive channels.

The list of ostensive cues included eye contact, lifting their eyebrows, pointing or referencing an object in space, and changing their tone of voice. And then Cecilia discovered another unexpected factor. We saw that the children, when they taught, would move around and get out of their chairs. We, from our place as researchers, would ask them to sit down to avoid distractions that would make it harder for us to detect their ostensive gestures. And, as we only realized later, that led us to miss the chance to make a discovery. When we didn't try to maintain order and we just let things follow their natural course, we discovered that the children, invariably, would stand up when they were teaching. Not one of them was sitting. They would get up and start moving around. We still have to discern whether that has to do with an ostensive gesture to mark the flow of knowledge; in other words: 'I am standing because I am the one who knows,' or if, rather, it's related to a question of irrepressible excitement caused by the rush of teaching.

In one of the experiments that Cecilia did, the children – between two and seven years old – had to teach an adult a very simple rule. A monkey was smelling flowers, and they had to find out which ones made the monkey sneeze. The only difficulty was that the flowers weren't always presented one by one so the game involved some deductive reasoning. But the game was simple enough for a two-year-old to easily solve it. After that, an adult would come and solve it incorrectly. The children thought that this was very funny. In fact, feigning incomprehension is a typical game between adults and children.

Most of the children responded by teaching the adult the tools needed to solve the problem. But a few of them said something like this: 'When they show you a flower, look at me. If it's the one that makes the monkey sneeze, I'll wink. And if it isn't, I'll lift my eyebrow.' They were cheating by offering to tell them the answer. On one hand, this shows the origins of this kind of copying in a school setting. But it also suggests something profound and central to teaching. Teachers of all types have to stop the class they are teaching from time to time if they feel that the students are not prepared for it. Where, when and how to do that is one of the most delicate problems in pedagogy. In a way, those seven-year-olds solved it by proposing a solution based on trick signals instead of explaining it. This suggests that if adults were incapable of doing something so simple, the children felt it was not worth trying to teach them, so they abandoned the pedagogy.*

Spikes of culture

Through exploring when and what we teach, we discovered that in infancy we were voracious, enthusiastic and effective teachers. But we still have to answer the toughest question: why do we teach? Why do we invest time and effort in sharing our knowledge with others? The *why* behind human behaviour almost always raises countless questions and unexamined answers.

Let's look at an apparently much simpler example: why do we drink water? We can give a utilitarian response: the body needs water in order to function. But no one drinks water because they understand that premise; we do it because we are thirsty. But, then, why do we get thirsty? Where does that desire to get up and seek out water come from? We can propose a reply from a biological perspective; in the brain there is a circuit which, when it detects that the body is dehydrated, links the motivation engine (dopamine) with water. But this only shifts the question: why do we have that circuit? And this avalanche of questions always ends in an argument about evolutionary history. If that mechanism weren't there and we didn't feel the desire to drink when our bodies lacked water, we would die of thirst. And, therefore, we wouldn't be here today, asking these questions.

But a system forged in the evolutionary kitchen is neither precise nor perfect. We like some things that are bad for us and we dislike some things that are good for us. Besides, the context changes, so that the same circuits that were functional at one point in evolutionary history cease to be so in another. For example, eating past the necessary

---

* Ironically, 'teaching' and 'cheating' are anagrams.

levels could be adaptive to stockpile calories in a period of shortage. But the same mechanism is harmful and becomes the driving force of addictions and obesity when there is, as often happens today, a larder filled with food. Thus a reasonable premise for understanding the genesis of the cerebral circuits that make us do what we do and be who we are is that in some contexts – not necessarily the current one – it was adaptive. It is an evolutionary view of the history of biological development.

These arguments can also be proposed, although not as firmly, to understand the propensity to behaviours that forge social being and culture. In this case – why it may be or may have been adaptive to teach – we can sketch out the following argument, which is best located in a simpler time than contemporary society: teaching other people to defend themselves from a predator is a way of protecting oneself. In the jungle, many non-human primates have a rudimentary language based on calls that warn of different dangers, such as snakes, eagles, big cats. Each danger has a different call. We can think of this as something analogous to the prelude to teaching in babies, an *argumentum ornitologicum*: a bird in a privileged position to see something that others do not will share that knowledge in a public message (a tweet). The fact that every bird has this instinct results in a collective alarm system that functions very well for the flock as a whole.

Sharing knowledge can be detrimental to the one who shares it (which in commercial terms is the reason behind all patents and the secret formula for Coca-Cola, for example). But we understand that, in many circumstances, disseminating information can create groups with resources that confer an advantage on the individuals who make up that group. These are, generally, the typical arguments for understanding the evolution of altruistic behaviours and a utilitarian reason for understanding the genesis of human communication. Teaching others is a way of taking care of ourselves.

The propensity to share knowledge is an individual trait that makes us invariably gather into groups. It is the seed of culture. Setting up cultural networks in small groups, tribes and collectives makes each individual function a bit better than they would alone. Beyond this utilitarian vision, teaching is also a way of getting to know not only things and causes but other people as well as ourselves.

Docendo discimus

Teaching is an intentional behaviour through which a teacher bridges a gap in knowledge. This compact definition presupposes many requirements in cognitive machinery that enables us to teach and to learn. For example:

(1) Recognizing our knowledge of something (metacognition). Recognizing the knowledge someone else has of something (theory of mind).
(2) Understanding that there is a disparity between these two sets of knowledge.
(3) Having the motivation to bridge that gap.
(4) Having a communicational apparatus (language, gestures) in order to bridge it.

Now, I propose a radical hypothesis about the first two points that comprise teaching, which naturally derive from the idea of the teaching instinct.

My conjecture is that children begin to teach as if compelled to do so, without taking into consideration what the student really knows or even what they themselves know. They could, in fact, teach a doll, the sea or a stone. From this point of view, teaching precedes – and can provide the experience for – forging a theory of mind. Teaching helps to put oneself mentally in another's shoes and to be able to attribute thoughts and intentions to others. In the same way, children teach things they aren't fully knowledgeable of and, in doing so, consolidate their own knowledge. This is a way of revisiting and delving deeper into Seneca's celebrated idea: *Docendo discimus* – through teaching, we learn. We not only learn about what we are teaching but we also learn to calibrate our own and others' knowledge. In addition to becoming more versed in the subject, when we teach we also learn about ourselves and others.

We saw that learning is about expressing new information in the framework of the language of individual thinking. Teaching is an exercise in translation in which we learn not only because we review facts – hit the books, as we say – but because we carry out the exercise of simplifying, summing up, underlining and thinking about how the same problem is seen from another's perspective. All these tasks, so intrinsic to pedagogy, are the essential fuel of learning.

Someone with a well-consolidated grasp of the theory of mind can reflect from another's perspective and thus understand that two people can come to different conclusions. This can be demonstrated in the laboratory in the following way. The first person sees a packet of sweets. There is no way of seeing what is inside it. They also see how someone takes out all the sweets and puts screws in instead. Then Bill, who hasn't seen any of this, comes in. The question for the first person is: what does Bill think is inside the packet? In order to respond, the first person must travel to the other's thoughts.

Someone equipped with a theory of mind understands that, from that perspective, the most natural thing for Bill to think is that the packet is filled with sweets. Someone who does not have a well-established theory of mind supposes that Bill thinks there must be screws inside. This simple example serves for a wide range of problems that include understanding that the other person not only has a body of knowledge that is different from yours but also another affective perspective, with other sensibilities and ways of reasoning. The theory of mind is expressed rudimentarily in the first months of life and then is slowly consolidated during development.

Cecilia Calero and I corroborated the first part of the hypothesis of learning as a process of consolidating the theory of mind. We saw that children didn't need to have calibrated a theory of others' knowledge in order to teach. Children teach even when they barely have any idea of what the other person knows. What we still must discover, by carefully following the development of those little teachers, is whether the most interesting hypothesis is true: if, when teaching, the children forge and consolidate the theory of mind.

The second hypothesis of the *teaching instinct* – teaching helps to consolidate the knowledge of the one teaching – today has a far wider consensus. Seneca's baton was picked up by Joseph Joubert, the inspector-general of universities under Napoleon, with his famous phrase: 'To teach is to learn twice.' And the contemporary version of this idea – according to which one way of learning is by sometimes putting yourself in the place of the teacher – begins with a concrete and practical necessity of our educational system. Assigning tutors to students is the most effective educational intervention. But assigning an expert tutor to each student is completely implausible. One solution that has been tested successfully in many innovative educational systems is peer tutoring, students who temporarily assume the role of teachers in order to complement their classmates' education. This happens spontaneously in rural schools, where there are few students, of varying ages, who share the same classroom. It also happens, naturally, outside the school environment.

Andrea Moro, one of the greatest contemporary linguists, noticed that children's mother tongue is not their mother's language but that of their friends. Children who grow up in a foreign country speak their peers' language more naturally than their parents' tongue. Bringing peer tutoring to the classroom is simply installing in formal education something common and effective in *the school of life*.

Even if peer teaching is not as effective as expert tutoring, it has a great advantage above and beyond practical and economic considerations. The tutor also learns while teaching. This effect is observed even when the tutor and student are the same age and even if the teaching is reciprocal, meaning the children alternate their teaching and learning roles.

This is promising and should encourage the practice in educational settings. But there is an important caveat: the effect is highly variable. In some cases, the children improve greatly as they teach. In other cases, they don't. If we understood when this practice is useful, we would have an effective recipe for improving education and, along the way, we would have revealed an important secret about learning.

That is what Rod Roscoe and Michelene Chi did discovering that tutors benefit more from their teaching when it fulfils these principles:

(1) The teachers rehearse and put their knowledge to the test, which allows them to detect errors, bridge gaps and generate new ideas.

(2) The teachers establish analogies or metaphors, relating the different concepts and assigning priorities to the information they have. Teaching is not listing facts but rather constructing a story that links them together in a plot.

These principles are very similar to a concept we have already looked at, the memory palace. The construction of memory is more similar to a creative process than to a passive storage of information in nooks and crannies of the brain. The memories become effective, strong and long lasting if they are reorganized into a reasonable visual plot, with a certain logic to the palace's architectural structure. Now we can extend this idea to all thought. Students, when teaching, are organizing concepts that they've already acquired into a new architecture that is more propitious to remembering and, above all, to the construction of new knowledge. They are building their palaces of thought.

EPILOGUE

I was about sixteen years old and had just read a very short story that told of a couple who loved each other as intensely as two people could. One evening they make glorious love to each other and then he goes to take a shower. She is smoking in bed, as she savours the love lingering in her body. He has an unexpected, tragic fall, tripping and hitting his head against the bathtub. He dies in silence, without anyone, not even her, realizing it has happened. The story is about that second in which they are barely three feet away from each other, while she is infinitely happy because of the love she feels for him and he is dead. I don't remember who wrote the story, or the title, just the cheap paper and bad printing of the magazine. Then I came across this same idea in the final story of an anthology edited by Borges and Bioy Casares, entitled 'The World is Wide and Strange': 'They say that Dante, in Chapter 40 of *La Vita Nuova*, says that when travelling through the streets of Florence he was surprised to find pilgrims who knew nothing of his beloved Beatrice.'

This book, and perhaps my whole adventure in science, is a way of answering the questions that hover implicit in those texts. I suspect that, in one way or another, we all share that impulse. That is the *raison d'être* of words, hugs, loves. As well as of quarrels, disputes, jealousy. Our feelings, our beliefs, our ideas are all expressed through the body's rudimentary language.

If I were to sum up the idea behind this book in one sentence, it would be the quest to make human thought transparent. From the first page to the last, the search for that transparency is a constant. All of these experiments with babies are designed to better comprehend their desires, needs and virtues, when their lack of language makes them opaque. Understanding how we make decisions, the driving force behind boldness, the reasons for our whims and our beliefs, is a way of removing a layer of opacity from thought itself, which is sometimes hidden beneath the mask of consciousness. And, finally, the pedagogy that is so prominent in the book's last chapter is, in my view of neuroscience, a human achievement that allows us to come together, to share what we know and what we think. So that the world is less wide and strange.

APPENDIX

The geography of the brain

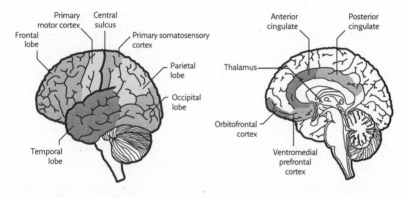

It is useful, when studying the brain, to divide it into different regions. Some of them are delineated by grooves or clefts. Using this method, the cerebral cortex, which includes the entire surface of the cerebral hemispheres, can be divided into four large regions: frontal, parietal, occipital and temporal. The parietal and frontal cortices, for example, are separated by the central sulcus. Each of these large regions of the cortex takes part in multiple functions but, at the same time, has a certain degree of specialization. The frontal cortex functions as the brain's 'control tower'. Without it, for example, we would be unable to refrain from eating in a situation in which we know it is not good for us (although we are hungry). The frontal cortex regulates, inhibits and operates different cerebral processes, and makes plans. The occipital cortex manages visual perception. The parietal cortex integrates and coordinates sensorial information with actions. It accounts for our ability to catch a ball by guiding our movements using in-real-time information about its speed and its trajectory. And the temporal cortex encodes memories, and works as a bridge between vision, hearing and language.

These large regions are in turn divided according to anatomical criteria or functional roles. For example, the motor cortex is the area of the frontal cortex that manages the muscles, and the somatosensory cortex is the area in the parietal cortex that coordinates tactile perception.

In the fissure that separates the two hemispheres of the brain, we can identify subdivisions of the frontal cortex. For example, the ventromedial prefrontal cortex and the orbitofrontal cortex are involved in different aspects of decision-making such as encoding the value and the risk of different options. The cingulate cortex extends beneath the frontal and parietal cortices. The section closest to the forehead (anterior cingulate) plays a primordial role in the ability to monitor and control our actions. For instance, when you realize you have made a mistake just after making an action, it is because the anterior cingulate has signalled it. On the other hand, the posterior cingulate (the part closest to the nape) activates when your mind wanders or when you are daydreaming. Finally, in the centre of the brain is the thalamus, which turns off when you are asleep or under anaesthesia, and it turns on when you wake up.

BIBLIOGRAPHY

1: The origin of thought

Abrams, David S., Marianne Bertrand and Sendhil Mullainathan. 'Do judges vary in their treatment of race?', *The Journal of Legal Studies* 41.2 (2012): 347–83

Anderson, Peter. 'Assessment and development of executive function (EF) during childhood', *Child Neuropsychology* 8.2 (2002): 71–82

Bialystok, Ellen. *Bilingualism in Development: Language, Literacy, and Cognition.* Cambridge University Press, 2001

Carey, Susan. *The Origin of Concepts.* Oxford University Press, 2009

Costa, Albert, Mireia Hernández and Núria Sebastián-Gallés. 'Bilingualism aids conflict resolution: evidence from the ANT task', *Cognition* 106.1 (2008): 59–86

Diamond, Adele. 'Executive functions', *Annual Review of Psychology* 64 (2013): 135–68

Diamond, Adele, and Patricia S. Goldman-Rakic. 'Comparison of human infants and rhesus monkeys on Piaget's AB task: evidence for dependence on dorsolateral prefrontal cortex', *Experimental Brain Research* 74.1 (1989): 24–40

Dipboye, Robert L., Richard D. Arvey and David E. Terpstra. 'Sex and physical attractiveness of raters and applicants as determinants of resume evaluations', *Journal of Applied Psychology* 62.3 (1977): 288

Gallese, Vittorio, and Alvin Goldman. 'Mirror neurons and the simulation theory of mind-reading', *Trends in Cognitive Sciences* 2.12 (1998): 493–501

Garbin, Gabrielle, et al. 'Bridging language and attention: brain basis of the impact of bilingualism on cognitive control', *Neuroimage* 53.4 (2010): 1272–8

Gergely, György, and Gergely Csibra. 'Teleological reasoning in infancy: the naïve theory of rational action', *Trends in Cognitive Sciences* 7.7 (2003): 287–92

Gergely, György, Harold Bekkering and Ildikó Király. 'Developmental psychology: rational imitation in preverbal infants', *Nature* 415.6873 (2002): 755

Hamlin, J. Kiley, and Karen Wynn. 'Young infants prefer prosocial to antisocial others', *Cognitive Development* 26.1 (2011): 30–39

Hamlin, J. Kiley, Karen Wynn and Paul Bloom. 'Social evaluation by preverbal infants', *Nature* 450.7169 (2007): 557–9

Hamlin, J. Kiley, et al. 'Not like me = bad: infants prefer those who harm dissimilar others', *Psychological Science* (2013): 589–94

Hood, Bruce, Susan Carey and Sandeep Prasada. 'Predicting the outcomes of physical events: two-year-olds fail to reveal knowledge of solidity and support', *Child Development* 71.6 (2000): 1540–54

Izard, Véronique, et al. 'Newborn infants perceive abstract numbers', *Proceedings of the National Academy of Sciences* 106.25 (2009): 10382–5

Kinzler, Katherine D., Emmanuel Dupoux and Elizabeth S. Spelke. 'The native language of social cognition', *Proceedings of the National Academy of Sciences* 104.30 (2007): 12577–80

Kuhl, Patricia K. 'Early language acquisition: cracking the speech code', *Nature Reviews Neuroscience* 5.11 (2004): 831–43

Meltzoff, Andrew N., and M. Keith Moore. 'Imitation of facial and manual gestures by human neonates', *Science* 198.4312 (1977): 75–8

Meltzoff, Andrew N., and Richard W. Borton. 'Intermodal matching by human neonates', *Nature* 282.5737 (1979): 403–4

Mesz, Bruno, Marcos A. Trevisan and Mariano Sigman. 'The taste of music', *Perception* 40.2 (2011): 209–19

Núñez, Rafael E., and Eve Sweetser. 'With the future behind them: convergent evidence from Aymara language and gesture in the crosslinguistic comparison of spatial construals of time', *Cognitive Science* 30.3 (2006): 401–50

Posner, Michael I., and Stanislas Dehaene. 'Attentional networks', *Trends in Neurosciences* 17.2 (1994): 75–9

Ramachandran, Vilayanur S., and Edward M. Hubbard. 'Synaesthesia – a window into perception, thought and language', *Journal of Consciousness Studies* 8.12 (2001): 3–34

Ramus, Franck, et al. 'Language discrimination by human newborns and by cotton-top tamarin monkeys', *Science* 288.5464 (2000): 349–51

Rochat, Philippe. *Early Social Cognition: Understanding Others in the First Months of Life*. Psychology Press, 2014

Rueda, M. Rosario, et al. 'Development of attentional networks in childhood', *Neuropsychologia* 42.8 (2004): 1029–40

Saffran, Jenny R., Richard N. Aslin and Elissa L. Newport. 'Statistical learning by 8-month-old infants', *Science* 274.5294 (1996) 1926–8

Schulz, Laura, Tamar Kushnir and Alison Gopnik. 'Learning from doing: intervention and causal inference', *Causal Learning: Psychology, Philosophy, and Computation* (2007): 67–85

Sigman, Mariano, et al. 'Neuroscience and education: prime time to build the bridge', *Nature Neuroscience* 17.4 (2014): 497–502

Valvo, Alberto. *Sight Restoration after Long Term Blindness: The Problems and Behavior Patterns of Visual Rehabilitation*. American Foundation for the Blind, 1971

2: The fuzzy borders of identity

Ais, Joaquín, et al. 'Individual consistency in the accuracy and distribution of confidence judgments', *Cognition* 146 (2016): 377–86

Ariely, Dan, and George Loewenstein. 'The heat of the moment: the effect of sexual arousal on sexual decision making', *Journal of Behavioral Decision Making* 19.2 (2006): 87–98

Barttfeld, Pablo, et al. 'Distinct patterns of functional brain connectivity correlate with objective performance and subjective beliefs', *Proceedings of the National Academy of Sciences* 110.28 (2013): 11577–82

Bechara, Antoine, et al. 'Deciding advantageously before knowing the advantageous strategy', *Science* 275.5304 (1997): 1293–5

Brodsky, Warren. *Driving With Music: Cognitive-Behavioural Implications*. Ashgate Publishing, 2015

Cesarini, David, et al. 'Heritability of cooperative behavior in the trust game', *Proceedings of the National Academy of Sciences* 105.10 (2008): 3721–6

Chaim, and Uri Gneezy. 'Discrimination in a segmented society: an experimental approach', *The Quarterly Journal of Economics* 116.1 (2001): 351–77

Charness, Gary, and Uri Gneezy. 'What's in a name? Anonymity and social distance in dictator and ultimatum games', *Journal of Economic Behavior & Organization* 68.1 (2008): 29–35

Costa, Albert, et al. 'Your morals depend on language', *PloS One* 9.4 (2014): e94842

Croson, Rachel, and Nancy Buchan. 'Gender and culture: international experimental evidence from trust games', *The American Economic Review* 89.2 (1999): 386–91

Croson, Rachel, and Uri Gneezy. 'Gender differences in preferences', *Journal of Economic Literature* 47.2 (2009): 448–74

Danziger, Shai, Jonathan Levav and Liora Avnaim-Pesso. 'Extraneous factors in judicial decisions', *Proceedings of the National Academy of Sciences* 108.17 (2011): 6889–92

Dehaene, Stanislas, et al. 'Imaging unconscious semantic priming', *Nature* 395.6702 (1998): 597–600

Delgado, Mauricio R., Robert H. Frank and Elisabeth A. Phelps. 'Perceptions of moral character modulate the neural systems of reward during the trust game', *Nature Neuroscience* 8.11 (2005): 1611–18

Di Tella, Rafael, et al. 'Conveniently upset: avoiding altruism by distorting beliefs about others' altruism', *The American Economic Review* 105.11 (2015): 3416–42

Dijksterhuis, Ap, et al. 'On making the right choice: the deliberation-without-attention effect', *Science* 311.5763 (2006): 1005–7

Drugowitsch, Jan, et al. 'Tuning the speed–accuracy trade-off to maximize reward rate in multisensory decision-making', *Elife* 4 (2015): e06678

Fleming, Stephen M., et al. 'Relating introspective accuracy to individual differences in brain structure', *Science* 329.5998 (2010): 1541–3

Gold, Joshua I., and Michael N. Shadlen. 'Banburismus and the brain: decoding the relationship between sensory stimuli, decisions, and reward', *Neuron* 36.2 (2002): 299–308

Graziano, Martin, and Mariano Sigman. 'The spatial and temporal construction of confidence in the visual scene', *PLoS One* 4.3 (2009): e4909

Greene, Joshua, and Jonathan Haidt. 'How (and where) does moral judgment work?', *Trends in Cognitive Sciences* 6.12 (2002): 517–23

Greene, Joshua D., et al. 'An fMRI investigation of emotional engagement in moral judgment', *Science* 293.5537 (2001): 2105–8

Guastella, Adam J., and Ian B. Hickie. 'Oxytocin treatment, circuitry, and autism: a critical review of the literature placing oxytocin into the autism context', *Biological Psychiatry* 79.3 (2016): 234–42

Haley, Kevin J., and Daniel M. T. Fessler. 'Nobody's watching?: Subtle cues affect generosity in an anonymous economic game', *Evolution and Human Behavior* 26.3 (2005): 245–56

Hall, Joshua C. *Homer Economicus: The Simpsons and Economics.* Stanford University Press, 2014

Hanks, Timothy D., Jochen Ditterich and Michael N. Shadlen. 'Micro-stimulation of macaque area LIP affects decision-making in a motion discrimination task', *Nature Neuroscience* 9.5 (2006): 682–9

Henrich, Joseph, et al. 'In search of homo economicus: behavioral experiments in 15 small-scale societies', *The American Economic Review* 91.2 (2001): 73–8

Johnson, Eric J., and Daniel Goldstein. 'Medicine. Do defaults save lives?', *Science* 302.5649 (2003): 1338–9

Kahneman, Daniel. *Thinking, Fast and Slow*. Macmillan, 2011

Kosfeld, Michael, et al. 'Oxytocin increases trust in humans', *Nature* 435.7042 (2005): 673–6

Leone, María J., et al. 'The tell-tale heart: heart rate fluctuations index objective and subjective events during a game of chess', *Frontiers in Human Neuroscience* (2012): doi 10.3389/fnhum. 2012.00273

Miller, Greg. 'The promise and perils of oxytocin', *Science* 339.6117 (2013): 267–9

Pedersen, Cort A., et al. 'Oxytocin induces maternal behavior in virgin female rats', *Science* 216.4546 (1982): 648–50

Reyna, Valerie F., and Frank Farley. 'Risk and rationality in adolescent decision making: implications for theory, practice, and public policy', *Psychological Science in the Public Interest* 7.1 (2006): 1–44

Rosenblat, Tanya S. 'The beauty premium: physical attractiveness and gender in dictator games', *Negotiation Journal* 24.4 (2008): 465–81

Salles, Alejo, et al. 'The metacognitive abilities of children and adults', *Cognitive Development* 40 (2016): 101–10

Shadlen, Michael N. and William T. Newsome. 'Neural basis of a perceptual decision in the parietal cortex (area LIP) of the rhesus monkey', *Journal of Neurophysiology* 86.4 (2001): 1916–36

Sharot, Tali, Christoph W. Korn and Raymond J. Dolan. 'How unrealistic optimism is maintained in the face of reality', *Nature Neuroscience* 14.11 (2011): 1475–9

Sharot, Tali, et al. 'Neural mechanisms mediating optimism bias', *Nature* 450.7166 (2007): 102–5

Strack, Fritz, Leonard L. Martin and Sabine Stepper. 'Inhibiting and facilitating conditions of the human smile: a nonobtrusive test of the facial feedback hypothesis', *Journal of Personality and Social Psychology* 54.5 (1988): 768

Thorndike, Edward L. 'A constant error in psychological ratings', *Journal of Applied Psychology* 4.1 (1920): 25–9

Todorov, Alexander, et al. 'Inferences of competence from faces predict election outcomes', *Science* 308.5728 (2005): 1623–6

Wedekind, Claus, et al. 'MHC-dependent mate preferences in humans', *Proceedings of the Royal Society of London B: Biological Sciences* 260.1359 (1995): 245–9

Young, Liane, et al. 'Disruption of the right temporoparietal junction with transcranial magnetic stimulation reduces the role of beliefs in moral judgments', *Proceedings of the National Academy of Sciences* 107.15 (2010): 6753–8

Zylberberg, Ariel, Pablo Barttfeld and Mariano Sigman. 'The construction of confidence in a perceptual decision', *Frontiers in Integrative Neuroscience* 6 (2012): 79

Zylberberg, Ariel, Pieter R. Roelfsema and Mariano Sigman. 'Variance misperception explains illusions of confidence in simple perceptual decisions', *Consciousness and Cognition* 27 (2014): 246–53

3: The machine that constructs reality

Baars, Bernard J. 'In the theatre of consciousness. Global Workspace Theory, a rigorous scientific theory of consciousness', *Journal of Consciousness Studies* 4.4 (1997): 292–309

Bekinschtein, Tristan A., et al. 'Classical conditioning in the vegetative and minimally conscious state', *Nature Neuroscience* 12.10 (2009): 1343–9

Blakemore, Sarah-Jayne, Daniel M. Wolpert and Chris D. Frith. 'Central cancellation of self-produced tickle sensation', *Nature Neuroscience* 1.7 (1998): 635–40

Blakemore, Sarah-Jayne, Daniel Wolpert and Chris Frith. 'Why can't you tickle yourself?', *Neuroreport* 11.11 (2000): R11–R16

Dehaene, Stanislas, and Lionel Naccache. 'Towards a cognitive neuroscience of consciousness: basic evidence and a workspace framework', *Cognition* 79.1 (2001): 1–37

Dehaene, Stanislas, et al. 'Conscious, preconscious, and subliminal processing: a testable taxonomy', *Trends in Cognitive Sciences* 10.5 (2006): 204–11

Dennett, Daniel C. *Consciousness Explained*. Penguin Books, 1993

Duhamel, Jean-René, Carol L. Colby and Michael E. Goldberg. 'The updating of the representation of visual space in parietal cortex by intended eye movements', *Science* 255.5040 (1992): 90

Edelman, Gerald M. 'Naturalizing consciousness: a theoretical framework', *Proceedings of the National Academy of Sciences* 100.9 (2003): 5520–24

Ford, Judith M., et al. 'Neurophysiological evidence of corollary discharge dysfunction in schizophrenia', *American Journal of Psychiatry* 158.12 (2001): 2069–71

Ford, Judith M., et al. 'Dissecting corollary discharge dysfunction in schizophrenia', *Psychophysiology* 44.4 (2007): 522–9

Fries, Pascal, et al. 'Modulation of oscillatory neuronal synchronization by selective visual attention', *Science* 291.5508 (2001): 1560–63

Frith, Chris. 'The self in action: lessons from delusions of control', *Consciousness and Cognition* 14.4 (2005): 752–70

Gazzaniga, Michael S. 'Principles of human brain organization derived from split-brain studies', *Neuron* 14.2 (1995): 217–228

Haynes, John-Dylan, and Geraint Rees. 'Predicting the orientation of invisible stimuli from activity in human primary visual cortex', *Nature Neuroscience* 8.5 (2005): 686–91

Johansson, Petter, et al. 'Failure to detect mismatches between intention and outcome in a simple decision task', *Science* 310.5745 (2005): 116–19

Kouider, Sid, et al. 'A neural marker of perceptual consciousness in infants', *Science* 340.6130 (2013): 376–80

Libet, Benjamin. 'Unconscious cerebral initiative and the role of conscious will in voluntary action', *Neurophysiology of Consciousness* (1993): 269–306

Libet, Benjamin, et al. 'Time of conscious intention to act in relation to onset of cerebral activity (readiness-potential)', *Brain* 106.3 (1983): 623–42

Monti, Martin M., et al. 'Willful modulation of brain activity in disorders of consciousness', *New England Journal of Medicine* 362.7 (2010): 579–89

Owen, Adrian M., et al. 'Detecting awareness in the vegetative state', *Science* 313.5792 (2006): 1402

Ross, John, et al. 'Changes in visual perception at the time of saccades', *Trends in Neurosciences* 24.2 (2001): 113–21

Shalom, Diego E., et al. 'Choosing in freedom or forced to choose? Introspective blindness to psychological forcing in stage-magic', *PloS One* 8.3 (2013): e58254

Sigman, Mariano, and Stanislas Dehaene. 'Parsing a cognitive task: a characterization of the mind's bottleneck', *PLoS Biol* 3.2 (2005): e37

Sommer, Marc A., and Robert H. Wurtz. 'Influence of the thalamus on spatial visual processing in frontal cortex', *Nature* 444.7117 (2006): 374–7

Soon, Chun Siong, et al. 'Unconscious determinants of free decisions in the human brain', *Nature Neuroscience* 11.5 (2008): 543–5

Sperry, Roger W., Michael S. Gazzaniga and Joseph E. Bogen. 'Interhemispheric relationships: the neocortical commissures; syndromes of hemisphere disconnection', *Handbook of Clinical Neurology* 4 (1969): 273–90

Tononi, Giulio, and Gerald M. Edelman. 'Consciousness and complexity', *Science* 282.5395 (1998): 1846–51

Weiskrantz, L., J. Elliott and C. Darlington. 'Preliminary observations on tickling oneself', *Nature* 230.5296 (1971): 598–9

4: Voyages of consciousness (or consciousness tripping)

Barttfeld, Pablo, et al. 'Signature of consciousness in the dynamics of resting-state brain activity', *Proceedings of the National Academy of Sciences* 112.3 (2015): 887–92

Bedi, Gillinder, et al. 'Automated analysis of free speech predicts psychosis onset in high-risk youths', *npj Schizophrenia* 1 (2015): 15030

Chang, L., et al. 'Marijuana use is associated with a reorganized visual-attention network and cerebellar hypoactivation', *Brain* 129.5 (2006): 1096–1112

Dave, Amish S., and Daniel Margoliash. 'Song replay during sleep and computational rules for sensorimotor vocal learning', *Science* 290.5492 (2000): 812–16

de Araujo, Draulio B., et al. 'Seeing with the eyes shut: neural basis of enhanced imagery following ayahuasca ingestion', *Human Brain Mapping* 33.11 (2012): 2550–60

Devane, William A., et al. 'Isolation and structure of a brain constituent that binds to the cannabinoid receptor', *Science* 258.5090 (1992): 1946–50

Diuk, Carlos, et al. 'A quantitative philology of introspection', *Frontiers in Integrative Neuroscience* 6 (2012): 80

Evers, Kathinka, and Mariano Sigman. 'Possibilities and limits of mind-reading: a neurophilosophical perspective', *Consciousness and Cognition* 22.3 (2013): 887–97

Fox, Michael D., et al. 'The human brain is intrinsically organized into dynamic, anticorrelated functional networks', *Proceedings of the National Academy of Sciences of the United States of America* 102.27 (2005): 9673–8

Fride, Ester, and Raphael Mechoulam. 'Pharmacological activity of the cannabinoid receptor agonist, anandamide, a brain constituent', *European Journal of Pharmacology* 231.2 (1993): 313–14

Grispoon, L.. *Marihuana Reconsidered*. Harvard University Press, 1971

Horikawa, Tomoyasu, et al. 'Neural decoding of visual imagery during sleep', *Science* 340.6132 (2013): 639–42

Kay, Kendrick N., et al. 'Identifying natural images from human brain activity', *Nature* 452.7185 (2008): 352–5

Killingsworth, Matthew A., and Daniel T. Gilbert. 'A wandering mind is an unhappy mind', *Science* 330.6006 (2010): 932

Koch, Marco, et al. 'Hypothalamic POMC neurons promote cannabinoid-induced feeding', *Nature* 519.7541 (2015): 45–50

Kouider, Sid, et al. 'Inducing task-relevant responses to speech in the sleeping brain', *Current Biology* 24.18 (2014): 2208–14

La Berge, Stephen P. 'Lucid dreaming as a learnable skill: a case study', *Perceptual and Motor Skills* 51.3_suppl2 (1980): 1039–42

Landauer, Thomas K., and Susan T. Dumais. 'A solution to Plato's problem: the latent semantic analysis theory of acquisition, induction, and representation of knowledge', *Psychological Review* 104.2 (1997): 211

Marshall, Lisa, et al. 'Boosting slow oscillations during sleep potentiates memory', *Nature* 444.7119 (2006): 610–13

Mechoulam, R., and Y. Gaoni. 'A total synthesis of dl–Δ1–tetrahydrocannabinol, the active constituent of hashish1', *Journal of the American Chemical Society* 87.14 (1965): 3273–5

Mooneyham, Benjamin W., and Jonathan W. Schooler. 'The costs and benefits of mind-wandering: a review', *Canadian Journal of Experimental Psychology/Revue canadienne de psychologie expérimentale* 67.1 (2013): 11

Nutt, David J. 'Equasy – an overlooked addiction with implications for the current debate on drug harms', *Journal of Psychopharmacology* 20 (2009): 3–5

Nutt, David J., Leslie A. King and Lawrence D. Phillips. 'Drug harms in the UK: a multicriteria decision analysis', *The Lancet* 376.9752 (2010): 1558–65

Purves, Dale, Joseph A. Paydarfar and Timothy J. Andrews. 'The wagon wheel illusion in movies and reality', *Proceedings of the National Academy of Sciences* 93.8 (1996): 3693–7

Reichle, Erik D., Andrew E. Reineberg and Jonathan W. Schooler. 'Eye movements during mindless reading', *Psychological Science* 21.9 (2010): 1300–1310

Ribeiro, Sidarta, and Miguel A. L. Nicolelis. 'Reverberation, storage, and postsynaptic propagation of memories during sleep', *Learning & Memory* 11.6 (2004): 686–96

Sigman, Mariano, and Guillermo A. Cecchi. 'Global organization of the Wordnet lexicon', *Proceedings of the National Academy of Sciences* 99.3 (2002): 1742–7

Stickgold, Robert. 'Sleep-dependent memory consolidation', *Nature* 437.7063 (2005): 1272–8

Tagliazucchi, Enzo, et al. 'Enhanced repertoire of brain dynamical states during the psychedelic experience', *Human Brain Mapping* 35.11 (2014): 5442–56

VanRullen, Rufin, Leila Reddy and Christof Koch. 'The continuous wagon wheel illusion is associated with changes in electroencephalogram power at 13 Hz', *Journal of Neuroscience* 26.2 (2006): 502–7

Voss, Ursula, et al. 'Lucid dreaming: a state of consciousness with features of both waking and non-lucid dreaming', *Sleep* 32.9 (2009): 1191–1200

Voss, Ursula, et al. 'Induction of self awareness in dreams through frontal low current stimulation of gamma activity', *Nature Neuroscience* 17.6 (2014): 810–12

Wagner, Ullrich, et al. 'Sleep inspires insight', *Nature* 427.6972 (2004): 352–5

Wotiz, John H., and Susanna Rudofsky. 'Kekulé's dreams: fact or fiction?', *Chemistry in Britain* 20 (1984): 720–23

Xie, Lulu, et al. 'Sleep drives metabolite clearance from the adult brain', *Science* 342.6156 (2013): 373–7

5: The brain is constantly transforming

Atkeson, Christopher G., et al. 'Using humanoid robots to study human behavior', *IEEE Intelligent Systems and Their Applications* 15.4 (2000): 46–56

Bao, Shaowen, Vincent T. Chan and Michael M. Merzenich. 'Cortical remodelling induced by activity of ventral tegmental dopamine neurons', *Nature* 412.6842 (2001): 79–83

Changizi, Mark A., and Shinsuke Shimojo. 'Character complexity and redundancy in writing systems over human history',

Proceedings of the Royal Society of London B: Biological Sciences 272.1560 (2005): 267–75

Chase, William G., and Herbert A. Simon. 'Perception in chess', *Cognitive Psychology* 4.1 (1973): 55–81

Cohen, Laurent, et al. 'Reading normal and degraded words: contribution of the dorsal and ventral visual pathways', *Neuroimage* 40.1 (2008): 353–66

Conway, Bevil R., and Doris Y. Tsao. 'Color-tuned neurons are spatially clustered according to color preference within alert macaque posterior inferior temporal cortex', *Proceedings of the National Academy of Sciences* 106.42 (2009): 18034–9

Dehaene, Stanislas. *Reading in the Brain: The New Science of How We Read*. Penguin Books, 2009

Dehaene, Stanislas, et al. 'The neural code for written words: a proposal', *Trends in Cognitive Sciences* 9.7 (2005): 335–41

Deutsch, Diana, et al. 'Absolute pitch among American and Chinese conservatory students: prevalence differences, and evidence for a speech-related critical period', *The Journal of the Acoustical Society of America* 119.2 (2006): 719–22

Ericsson, K. Anders, Ralf T. Krampe and Clemens Tesch-Römer. 'The role of deliberate practice in the acquisition of expert performance', *Psychological Review* 100.3 (1993): 363

Fehr, Ernst, and Urs Fischbacher. 'The nature of human altruism', *Nature* 425.6960 (2003): 785–91

Field, David J., Anthony Hayes and Robert F. Hess. 'Contour integration by the human visual system: evidence for a local "association field"', *Vision Research* 33.2 (1993): 173–93

Gilbert, Charles D., and Mariano Sigman. 'Brain states: top-down influences in sensory processing', *Neuron* 54.5 (2007): 677–96

Gilbert, Charles D., Mariano Sigman and Roy E. Crist. 'The neural basis of perceptual learning', *Neuron* 31.5 (2001): 681–97

Goldin, Andrea P., et al. 'From ancient Greece to modern education: universality and lack of generalization of the Socratic dialogue', *Mind, Brain, and Education* 5.4 (2011): 180–85

Goodale, Melvyn A., and A. David Milner. 'Separate visual pathways for perception and action', *Trends in Neurosciences* 15.1 (1992): 20–25

Holper, Lisa, et al. 'The teaching and the learning brain: a cortical hemodynamic marker of teacher–student interactions in the Socratic dialog', *International Journal of Educational Research* 59 (2013): 1–10

Hubel, David H., and Torsten N. Wiesel. 'Receptive fields, binocular interaction and functional architecture in the cat's visual cortex', *The Journal of Physiology* 160.1 (1962): 106–54

Lee, Tai Sing, and David Mumford. 'Hierarchical Bayesian inference in the visual cortex', *Journal of the Optical Society of America A* 20.7 (2003): 1434–48

Levitin, Daniel J., and Susan E. Rogers. 'Absolute pitch: perception, coding, and controversies', *Trends in Cognitive Sciences* 9.1 (2005): 26–33

Maguire, Eleanor A., et al. 'Routes to remembering: the brains behind superior memory', *Nature Neuroscience* 6.1 (2003): 90–95

McNamara, Danielle S., et al. 'Are good texts always better? Interactions of text coherence, background knowledge, and levels of understanding in learning from text', *Cognition and Instruction* 14.1 (1996): 1–43

O'Regan, J. Kevin. 'Eye movements and reading', *Reviews of Oculomotor Research* 4 (1990): 395

Shatz, Carla J. 'Emergence of order in visual system development', *Proceedings of the National Academy of Sciences* 93.2 (1996): 602–8

Sigman, M., and C. D. Gilbert. 'Learning to find a shape', *Nature Neuroscience* 3.3 (2000): 264–9

Sigman, Mariano, and Stanislas Dehaene. 'Brain mechanisms of serial and parallel processing during dual-task performance', *Journal of Neuroscience* 28.30 (2008): 7585–98

Sigman, Mariano, et al. 'On a common circle: natural scenes and Gestalt rules', *Proceedings of the National Academy of Sciences* 98.4 (2001): 1935–40

Sigman, Mariano, et al. 'Top-down reorganization of activity in the visual pathway after learning a shape identification task', *Neuron* 46.5 (2005): 823–35

Squire, Larry R. 'Memory and the hippocampus: a synthesis from findings with rats, monkeys, and humans', *Psychological Review* 99.2 (1992): 195

Thomas, Alexander, and Stella Chess. *Temperament and Development*. Brunner/Mazel, 1977

Todorov, Emanuel, and Michael I. Jordan. 'Optimal feedback control as a theory of motor coordination', *Nature Neuroscience* 5.11 (2002): 1226–35

Treisman, Anne M., and Garry Gelade. 'A feature-integration theory of attention', *Cognitive Psychology* 12.1 (1980): 97–136

Vinckier, Fabien, et al. '"What" and "where" in word reading: ventral coding of written words revealed by parietal atrophy', *Journal of Cognitive Neuroscience* 18.12 (2006): 1998–2012

Vinckier, Fabien, et al. 'Hierarchical coding of letter strings in the ventral stream: dissecting the inner organization of the visual word-form system', *Neuron* 55.1 (2007): 143–56

Zylberberg, Ariel, et al. 'The brain's router: a cortical network model of serial processing in the primate brain', *PLoS Comput Biol* 6.4 (2010): e1000765

6: Educated brains

Bonawitz, Elizabeth, et al. 'The double-edged sword of pedagogy: instruction limits spontaneous exploration and discovery', *Cognition* 120.3 (2011): 322–30

Bruer, John T. 'Education and the brain: a bridge too far', *Educational Researcher* 26.8 (1997): 4–16

Bus, Adriana G., and Marinus H. Van IJzendoorn. 'Phonological awareness and early reading: a meta-analysis of experimental training studies', *Journal of Educational Psychology* 91 (1999): 403–14

Calero, C. I., et al. 'Young children are natural pedagogues', *Cognitive Development* 35 (2015): 65–78

Church, R. Breckinridge, and Susan Goldin-Meadow. 'The mismatch between gesture and speech as an index of transitional knowledge', *Cognition* 23.1 (1986): 43–71

Cornell, James M. 'Spontaneous mirror-writing in children', *Canadian Journal of Psychology/Revue canadienne de psychologie* 39.1 (1985): 174

Csibra, Gergely, and György Gergely. 'Natural pedagogy', *Trends in Cognitive Sciences* 13.4 (2009): 148–53

Dehaene, Stanislas, et al. 'Core knowledge of geometry in an Amazonian indigene group', *Science* 311.5759 (2006): 381–4

Guttorm, Tomi K., et al. 'Brain event-related potentials (ERPs) measured at birth predict later language development in children with and without familial risk for dyslexia', *Cortex* 41.3 (2005): 291–303

Guttorm, Tomi K., et al. 'Newborn event-related potentials predict poorer pre-reading skills in children at risk for dyslexia', *Journal of Learning Disabilities* 43.5 (2010): 391–401

Liszkowski, Ulf, et al. '12- and 18-month-olds point to provide information for others', *Journal of Cognition and Development* 7.2 (2006): 173–87

Molfese, Dennis L. 'Predicting dyslexia at 8 years of age using neonatal brain responses', *Brain and Language* 72.3 (2000): 238–45

Pica, Pierre, et al. 'Exact and approximate arithmetic in an Amazonian indigene group', *Science* 306.5695 (2004): 499–503

Roscoe, Rod D., and Michelene T. H. Chi. 'Understanding tutor learning: knowledge-building and knowledge-telling in peer tutors' explanations and questions', *Review of Educational Research* 77.4 (2007): 534–74

Roscoe, Rod D., and Michelene T. H. Chi. 'Tutor learning: the role of explaining and responding to questions', *Instructional Science* 36.4 (2008): 321–50

Stokes, Donald E. *Pasteur's Quadrant: Basic Science and Technological Innovation*. Brookings Institution Press, 2011

Strauss, Sidney, Cecilia I. Calero and Mariano Sigman. 'Teaching, naturally', *Trends in Neuroscience and Education* 3.2 (2014): 38–43

Vosniadou, Stella, and William F. Brewer. 'Mental models of the earth: a study of conceptual change in childhood', *Cognitive Psychology* 24.4 (1992): 535–85

ACKNOWLEDGEMENTS

This book is the tale of a journey to the most hidden recesses of our brain and our thoughts, a journey of many years, which I undertook with friends, colleagues and travel companions both on the road and on the road of life.

I am infinitely grateful to all those in Argentina who accompanied me on the adventure of developing these ideas, and helped me build a profoundly interdisciplinary, provocative and plural space: all the doctoral and postdoctoral students and the researchers at the Integrative Neuroscience Laboratory at the Faculty of Natural and Exact Sciences at the University of Buenos Aires and the Neuroscience Laboratory at the Torcuato Di Tella University. I would also like to thank my colleagues and companions in New York and Paris, with whom these ideas continued to take shape. The concepts I discuss in this book were originally moulded with Gabriel Mindlin, Marcelo Magnasco, Charles Gilbert, Torsten Wiesel, Guillermo Cecchi, Michael Posner, Leopoldo Petreanu, Pablo Meyer Rojas, Eugenia Chiappe, Ramiro Freudenthal, Lucas Sigman, Martín Berón de Astrada, Stanislas Dehaene, Ghislaine Dehaene-Lambertz, Tristán Bekinschtein, Inés Samengo, Marcelo Rubinstein, Diego Golombek, Draulio Araujo, Kathinka Evers, Andrea P. Goldin, Cecilia Inés Calero, Diego Shalom, Diego Fernández Slezak, María Juliana Leone, Carlos Diuk, Ariel Zylberberg, Juan Frenkel, Pablo Barttfeld, Andrés Babino, Sidarta Ribeiro, Marcela Peña, David Klahr, Alejandro Maiche, Juan Valle

Lisboa, Jacques Mehler, Marina Nespor, Antonio Battro, Andrea Moro, Sidney Strauss, John Bruer, Susan Fitzpatrick, Marcos Trevisan, Sebastián Lipina, Bruno Mesz, Mariano Sardon, Horacio Sbaraglia, Albert Costa, Silvia Bunge, Jacobo Sitt, Andrés Rieznik, Gustavo Faigenbaum, Rafael Di Tella, Iván Reydel, Elizabeth Spelke, Susan Goldin-Meadow, Andrew Meltzoff, Manuel Carreiras, Michael Shadlen and John Duncan. I am grateful to my dad, for sharing his love and passion for psychiatry and for the study and comprehension of the human mind. My reading of Freud's works, with his handwritten annotations and underlining, lies at the heart of this project.

The foundations of this book have been in my brain – which according to its Spanish etymology means 'what's carried inside the head' – for years. But bringing it to fruition was an extraordinary adventure, and much more challenging and thrilling than I had imagined. And, of course, it would not have been possible without many people's support along the way. I want to thank them now, as I approach the finishing line. First of all, Florencia Ure and Roberto Montes, my publishers, who were at the start of this story. Roberto, in that first meeting – which seems an eternity ago – said in passing that the key was writing an honest book. Those casually spoken words resonated with me for a long time, like an anchor, as I gave shape to this project, and I endeavoured to do just that. Florencia Grieco accompanied me – both by my side and as the wind at my back – in the editing of the text. Countless meetings, emails, matés and coffees, with much back and forth during which I learned from her how to shape these ideas. Marcos Trevisan, my companion on so many adventures, saw me through this one with extraordinary patience. He taught me to read aloud and to think about the history behind words, and, above all, made me laugh during the most arduous stretches of writing. In the final sprint, Andre Goldin showed infinite generosity, enduring my giddy days and insomniac nights, as we went over both the science and the form of the book. Christián Carman revised historical and philosophical passages. Many thanks as well to the guys at El Gato y La Caja, Juan Manuel Garrido, Facundo Álvarez Heduan and Pablo

González, and to Andrés Rieznik, Cecilia Calero, Pablo Polosecki, Mercedes Dalessandro, Hugo Sigman, Silvia Gold, Juan Sigman and Claire Landmann, who read these pages and gave me their notes, observations and the occasional hug, which helped me to keep on rowing even when the wind was blowing hard.

In this book I have expressed the view of many scientists and philosophers who argue that language is not only a vehicle to convey thoughts, but also a tool to shape them. Even our moral judgements depend on the language in which they are presented. This book was first conceived in Spanish and then rewritten in English. I experienced this process as much more than a mere translation; it was a revision that was only possible by rethinking this book from the (quite imperfect) expressions of a distant language.

This journey from Spain and Latin America to the rest of the world began in New York and was made possible thanks to my agent, Max Brockman, to whom I am infinitely grateful. Then I had the enormous fortune to work closely with Mara Faye Lethem who translated the book to English and who, with Betina Gonzalez, an author I greatly admire, helped me with the final cut of *The Secret Life of the Mind*. I am honored to have this book published in the United States by Little, Brown and Company, and I am deeply thankful to Tracy Behar, my U.S. editor, and Ian Straus.

And from New York and Buenos Aires to London, to all my British companions at HarperCollins. Thanks to Lottie Fyfe, Katherine Patrick, Mark Handsley, and special thanks to Arabella Pike, my wonderful editor, for making me understand – or feel – the meaning of 'you'll never walk alone'.

INDEX